T0342189

Selfsimilar Processes

PRINCETON SERIES IN APPLIED MATHEMATICS

TITLES IN THE SERIES

Chaotic Transitions in Deterministic and Stochastic Dynamical Systems: Applications of Melnikov Processes in Engineering, Physics and Neuroscience by Emil Simiu

Selfsimilar Processes by Paul Embrechts and Makoto Maejima

Self-Regularity: A New Paradigm for Primal-Dual Interior Point Algorithms by Jiming Peng, Cornelis Roos and Tamás Terlaky

The Princeton Series in Applied Mathematics publishes high quality advanced texts and monographs in all areas of applied mathematics. Books include those of a theoretical and general nature as well as those dealing with the mathematics of specific applications areas and real-world situations.

Selfsimilar Processes

Paul Embrechts and Makoto Maejima

PRINCETON UNIVERSITY PRESS

PRINCETON AND OXFORD

Published by Princeton University Press,
41 William Street, Princeton, New Jersey 08540

In the United Kingdom: Princeton University Press,
3 Market Place, Woodstock, Oxfordshire OX20 1SY

Library of Congress Cataloging-in-Publication Data applied for.
Embrechts, Paul & Maejima, Makoto
Selfsimilar Processes/Paul Embrechts and Makoto Maejima
p.   cm.
Includes bibliographical references and index.
ISBN 0-691-09627-9 (alk. paper)

British Library Cataloging-in-Publication Data is available

www.pup.princeton.edu

"Voor mijn ouders. Hartelijk dank voor de liefde en de steun."

Paul Embrechts

Contents

Preface   ix

Chapter 1. Introduction   1

  1.1 Definition of Selfsimilarity   1
  1.2 Brownian Motion   4
  1.3 Fractional Brownian Motion   5
  1.4 Stable Lévy Processes   9
  1.5 Lamperti Transformation   11

Chapter 2. Some Historical Background   13

  2.1 Fundamental Limit Theorem   13
  2.2 Fixed Points of Renormalization Groups   15
  2.3 Limit Theorems (I)   16

Chapter 3. Selfsimilar Processes with Stationary Increments   19

  3.1 Simple Properties   19
  3.2 Long-Range Dependence (I)   21
  3.3 Selfsimilar Processes with Finite Variances   22
  3.4 Limit Theorems (II)   24
  3.5 Stable Processes   27
  3.6 Selfsimilar Processes with Infinite Variance   29
  3.7 Long-Range Dependence (II)   34
  3.8 Limit Theorems (III)   37

Chapter 4. Fractional Brownian Motion   43

  4.1 Sample Path Properties   43
  4.2 Fractional Brownian Motion for $H \neq 1/2$ is not a Semimartingale   45
  4.3 Stochastic Integrals with respect to Fractional Brownian Motion   47
  4.4 Selected Topics on Fractional Brownian Motion   51
     4.4.1 Distribution of the Maximum of Fractional Brownian Motion   51
     4.4.2 Occupation Time of Fractional Brownian Motion   52
     4.4.3 Multiple Points of Trajectories of Fractional Brownian Motion   53
     4.4.4 Large Increments of Fractional Brownian Motion   54

Chapter 5. Selfsimilar Processes with Independent Increments 57

  5.1 K. Sato's Theorem 57
  5.2 Getoor's Example 60
  5.3 Kawazu's Example 61
  5.4 A Gaussian Selfsimilar Process with Independent Increments 62

Chapter 6. Sample Path Properties of Selfsimilar Stable Processes with Stationary
         Increments 63

  6.1 Classification 63
  6.2 Local Time and Nowhere Differentiability 64

Chapter 7. Simulation of Selfsimilar Processes 67

  7.1 Some References 67
  7.2 Simulation of Stochastic Processes 67
  7.3 Simulating Lévy Jump Processes 69
  7.4 Simulating Fractional Brownian Motion 71
  7.5 Simulating General Selfsimilar Processes 77

Chapter 8. Statistical Estimation 81

  8.1 Heuristic Approaches 81
    8.1.1 The R/S-Statistic 82
    8.1.2 The Correlogram 85
    8.1.3 Least Squares Regression in the Spectral Domain 87
  8.2 Maximum Likelihood Methods 87
  8.3 Further Techniques 90

Chapter 9. Extensions 93

  9.1 Operator Selfsimilar Processes 93
  9.2 Semi-Selfsimilar Processes 95

References 101

Index 109

Preface

First, a word about the title "Selfsimilar Processes". Let there be no doubt, we all are very much in debt to Professor Benoit Mandelbrot. In his honor, we should have used as a title "Self-Affine Processes"; the notion of selfsimilarity, however, seems to have won the day and is used throughout an enormous literature. We therefore prefer to stick to it. In his most recent book [Man01], the author discusses these points in greater depth. Second, why this text at this point in time? As so often happens, the (re)emergence of scientific ideas related to a specific topic is like the sudden appearance of crocuses and daffodils in spring: the time is just right! Especially the availability of large data sets at ever finer and finer time resolution led to the increased analysis of stochastic processes at these fine time scales. Selfsimilar processes offer such a tool. We personally were motivated by recent applications in such fields as physics and mathematical finance. Researchers in those fields are increasingly interested in having a summary of the main results and guidance on existing literature. In addition to the numerous excellent papers existing on the subject, by far the best summary on the mathematics of the subject is to be found in [SamTaq94]. The history of selfsimilarity can without doubt best be learned from the various publications of Mandelbrot; see the list of references at the back. Mandelbrot's more recent publications contain a wealth of new ideas for researchers to look at, the notion of multifractability is just one example; see [Man99].

Our text should be viewed as intermediate lecture notes trying to bridge the gap between the various existing developments on the subject. The scientific community still awaits a definitive text. We hope that our contribution will be helpful for someone undertaking such an endeavor.

We both take great pleasure in thanking various colleagues and friends for helping with the presentation of this manuscript. Patrick Cheridito read various versions of the manuscript and gave us very valuable advice which resulted in numerous improvements. Hansruedi Künsch taught us some of the statistical issues. Andrea Binda produced the various figures. Mrs. Gabriele Baltes did an excellent job as overall technical editor, making sure that Japanese and Swiss versions of American LaTeX linked up. The second author gratefully acknowledges financial support from the ETH Forschungsinstitut für Mathematik, which allowed him to spend part of a sabbatical in Zürich.

Finally, both authors would like to thank their families for the constant support. Without their help, not only this text would not have been written, but also the numerous "get togethers" of both authors over so many years would not have been possible. Hence many thanks to Gerda, Krispijn, Eline, Frederik, Keiko and Utako.

Paul Embrechts, Zürich
Makoto Maejima, Yokohama

March, 2002

Selfsimilar Processes

Chapter One

Introduction

Selfsimilar processes are stochastic processes that are invariant in distribution under suitable scaling of time and space (see Definition 1.1.1).

It is well known that Brownian motion is selfsimilar (see Theorem 1.2.1). Fractional Brownian motion (see Section 1.3 and Chapter 4), which is a Gaussian selfsimilar process with stationary increments, was first discussed by Kolmogorov [Kol40]. The first paper giving a rigorous treatment of general selfsimilar processes is due to Lamperti [Lam62], where a fundamental limit theorem was proved (see Section 2.1). Later, the study of non-Gaussian selfsimilar processes with stationary increments was initiated by Taqqu [Taq75], who extended a non-Gaussian limit theorem by Rosenblatt [Ros61] (see Sections 2.3 and 3.4).

On the other hand, the works of Sinai [Sin76] and Dobrushin [Dob80] in the field of statistical physics, for instance, appeared around 1976 (see Section 2.2). It seems that similar problems were attacked independently in the fields of probability theory and statistical physics (see [Dob80]). The connection between these developments was made by Dobrushin. An early bibliographical guide is to be found in [Taq86].

1.1 DEFINITION OF SELFSIMILARITY

In the following, by $\{X(t)\} \overset{d}{=} \{Y(t)\}$, we denote equality of all joint distributions for $\mathbb{R}^d$-valued stochastic processes $\{X(t), t \geq 0\}$ and $\{Y(t), t \geq 0\}$ defined on some probability space $(\Omega, \mathcal{F}, P)$. Occasionally we simply write $X(t) \overset{d}{=} Y(t)$. Also $X(t) \overset{d}{=} Y(t)$ denotes equality of the marginal distributions for fixed t. By $X_n(t) \overset{d}{\Rightarrow} Y(t)$, we denote convergence of all joint distributions as $n \to \infty$, and by $\xi_n \overset{d}{\to} \xi$, the convergence in law of random variables $\{\xi_n\}$ to ξ. $\mathcal{L}(X)$ stands for the law of a random variable X. The characteristic function of a probability distribution μ is denoted by $\hat{\mu}(\theta)$, $\theta \in \mathbb{R}^d$. For $x \in \mathbb{R}^d$, $|x|$ is the Euclidean norm of x and x' is the transposed vector of x.

Definition 1.1.1 An $\mathbb{R}^d$-valued stochastic process $\{X(t), t \geq 0\}$ is said to be "selfsimilar" if for any $a > 0$, there exists $b > 0$ such that

$$\{X(at)\} \overset{d}{=} \{bX(t)\}. \tag{1.1.1}$$

We say that $\{X(t), t \geq 0\}$ is *stochastically continuous* at t if for any $\varepsilon > 0$, $\lim_{h \to 0} P\{|X(t + h) - X(t)| > \varepsilon\} = 0$. We also say that $\{X(t), t \geq 0\}$ is *trivial* if $X(t)$ is a constant almost surely for every t.

Theorem 1.1.1 [Lam62]   *If $\{X(t), t \geq 0\}$ is nontrivial, stochastically continuous at $t = 0$ and selfsimilar, then there exists a unique $H \geq 0$ such that b in (1.1.1) can be expressed as $b = a^H$.*

As there is some confusion about this result in the more applied literature, we prefer to give a proof. We start with an easy lemma.

Lemma 1.1.1   *If X is a nonzero random variable in $\mathbb{R}^d$, and if $b_1 X \overset{d}{\sim} b_2 X$ with $b_1, b_2 > 0$, then $b_1 = b_2$.*

Proof. Suppose $b_1 \neq b_2$. Then $X \overset{d}{\sim} bX$ with some $b \in (0,1)$. Hence $X \overset{d}{\sim} b^n X$ for any $n \in \mathbb{N}$, and, letting $n \to \infty$, we have $X = 0$ almost surely, which is a contradiction. $\square$

Proof of Theorem 1.1.1. Suppose $X(at) \overset{d}{\sim} b_1 X(t) \overset{d}{\sim} b_2 X(t)$. If $X(t)$ is nonzero for this t, then $b_1 = b_2$ by Lemma 1.1.1. By the nontriviality of $\{X(t)\}$, such a t exists. Thus $b_1 = b_2$, namely b in (1.1.1) is uniquely determined by a. We write $b = b(a)$. Then

$$X(aa't) \overset{d}{\sim} b(a)X(a't) \overset{d}{\sim} b(a)b(a')X(t).$$

Hence we have $b(aa') = b(a)b(a')$. We next show the monotonicity of $b(a)$. Suppose $a < 1$ and let $n \to \infty$ in $X(a^n) \overset{d}{\sim} b(a)^n X(1)$. Since $X(a^n)$ tends to $X(0)$ in probability by the stochastic continuity of $\{X(t)\}$ at $t = 0$, we must have that $b(a) \leq 1$. Since $b(a_1/a_2) = b(a_1)/b(a_2)$, if $a_1 < a_2$, then $b(a_1) \leq b(a_2)$, and thus $b(a)$ is nondecreasing. We have now concluded that $b(a)$ is nondecreasing and satisfies

$$b(aa') = b(a)b(a').$$

Thus $b(a) = a^H$ for some unique constant $H \geq 0$. $\square$

We call H the exponent of selfsimilarity of the process $\{X(t), t \geq 0\}$. We refer to such a process as H-selfsimilar (or H-ss, for short).

Property 1.1.1   *If $\{X(t), t \geq 0\}$ is H-ss and $H > 0$, then $X(0) = 0$ almost surely.*

Proof. By Definition 1.1.1, $X(0) \overset{d}{\sim} a^H X(0)$ and it is enough to let $a \to 0$. $\square$

Property 1.1.1 does not hold when $H = 0$.

Example 1.1.1 [Kon84] Let $\{Y(s), s \in \mathbb{R}\}$ be a strictly stationary process, ξ a random variable independent of $\{Y(s)\}$, and define $\{X(t), t \geq 0\}$ by

$$X(t) = \begin{cases} Y(\log t), & t > 0, \\ \xi, & t = 0. \end{cases}$$

For $t > 0$,

$$X(at) = Y(\log at) = Y(\log a + \log t)$$

$$\stackrel{d}{=} Y(\log t) = X(t)$$

so that

$$\{\{X(at), t > 0\}, \xi\} \stackrel{d}{=} \{\{X(t), t > 0\}, \xi\},$$

implying that $\{X(t), t \geq 0\}$ is 0-ss. However $X(0) \neq 0$. $\square$

Actually we have the following for $H = 0$.

Theorem 1.1.2 *Under the same assumptions of Theorem 1.1.1, $H = 0$ if and only if $X(t) = X(0)$ almost surely for every $t > 0$.*

Proof. The "if" part is trivial. For the "only if" part, by the property of 0-ss, $\{X(at)\} \stackrel{d}{=} \{X(t)\}$. Then for each $a > 0$, the joint distributions at $t = 0$ and $t = s/a$ are the same:

$$(X(0), X(s)) \stackrel{d}{\sim} (X(0), X(s/a)).$$

Hence for any $\varepsilon > 0$,

$$P\{|X(s) - X(0)| > \varepsilon\} = P\{|X(s/a) - X(0)| > \varepsilon\}.$$

The right-hand side of the above converges to 0 as $a \rightarrow \infty$, because of the stochastic continuity of the process at $t = 0$. Hence for each $s > 0$

$$P\{|X(s) - X(0)| > \varepsilon\} = 0, \qquad \forall \varepsilon > 0,$$

so that $X(s) = X(0)$ almost surely. $\square$

From the above considerations, it seems natural to consider only selfsimilar processes such that they are stochastically continuous at 0 and their exponents H are positive. Without further explicit mention, selfsimilarity will always be used in conjunction with $H > 0$ and stochastic continuity at 0.

1.2 BROWNIAN MOTION

An $\mathbb{R}^d$-valued stochastic process $\{X(t), t \geq 0\}$ is said to have *independent increments*, if for any $m \geq 1$ and for any partition $0 \leq t_0 < t_1 < \cdots < t_m$, $X(t_1) - X(t_0), \ldots, X(t_m) - X(t_{m-1})$ are independent, and is said to have *stationary increments*, if any joint distribution of $\{X(t + h) - X(h), \ t \geq 0\}$ is independent of $h \geq 0$. We will always use the term *stationarity* for the invariance of joint distributions under time shifts. Usually, this is referred to as strict stationarity. This is distinct from weak stationarity where time shift invariance is only required for the mean and covariance functions.

Definition 1.2.1 *If an $\mathbb{R}^d$-valued stochastic process $\{B(t), t \geq 0\}$ satisfies*

(a) $B(0) = 0$ *almost surely,*

(b) *it has independent and stationary increments,*

(c) *for each $t > 0$, $B(t)$ has a Gaussian distribution with mean zero and covariance matrix tI (where I is the identity matrix), and*

(d) *its sample paths are continuous almost surely,*

then it is called (standard) Brownian motion.

Theorem 1.2.1 *Brownian motion $\{B(t), t \geq 0\}$ is $\frac{1}{2}$-ss.*

Proof. It is enough to show that for every $a > 0$, $\{a^{-1/2}B(at)\}$ is also Brownian motion. Conditions (a), (b) and (d) follow from the same conditions for $\{B(t)\}$. As to (c), Gaussianity and the mean zero property also follow from the properties of $\{B(t)\}$. Moreover, $E[(a^{-1/2}B(at))(a^{-1/2}B(at))'] = tI$, thus $\{a^{-1/2}B(at)\}$ is Brownian motion. $\square$

Theorem 1.2.2 $E[B(t)B(s)'] = \min\{t, s\}I$.

Proof. We have

$E[B(t)B(s)']$

$$= \frac{1}{2}\{E[B(t)B(t)'] + E[B(s)B(s)'] - E[(B(t) - B(s))(B(t) - B(s))']\}$$

$$= \frac{1}{2}\{E[B(t)B(t)'] + E[B(s)B(s)'] - E[B(|t - s|)B(|t - s|)']\}$$

$$= \frac{1}{2}\{t + s - |t - s|\}I = \min\{t, s\}I. \quad \square$$

1.3 FRACTIONAL BROWNIAN MOTION

The following basic result for general selfsimilar processes with stationary increments leads to a natural definition of fractional Brownian motion.

Theorem 1.3.1 [Taq81] *Let* $\{X(t)\}$ *be real-valued H-selfsimilar with stationary increments and suppose that* $E[X(1)^2] < \infty$. *Then*

$$E[X(t)X(s)] = \frac{1}{2}\{t^{2H} + s^{2H} - |t - s|^{2H}\}E\Big[X(1)^2\Big].$$

Proof. By selfsimilarity and stationarity of the increments,

$$E[X(t)X(s)] = \frac{1}{2}\Big\{E\Big[X(t)^2\Big] + E\Big[X(s)^2\Big] - E\Big[(X(t) - X(s))^2\Big]\Big\}$$

$$= \frac{1}{2}\Big\{E\Big[X(t)^2\Big] + E\Big[X(s)^2\Big] - E\Big[X(|t - s|)^2\Big]\Big\}$$

$$= \frac{1}{2}\{t^{2H} + s^{2H} - |t - s|^{2H}\}E\Big[X(1)^2\Big]. \quad \square$$

Definition 1.3.1 *Let* $0 < H \le 1$. *A real-valued Gaussian process* $\{B_H(t),\ t \ge 0\}$ *is called "fractional Brownian motion" if* $E[B_H(t)] = 0$ *and*

$$E[B_H(t)B_H(s)] = \frac{1}{2}\{t^{2H} + s^{2H} - |t - s|^{2H}\}E\Big[B_H(1)^2\Big]. \qquad (1.3.1)$$

Remark 1.3.1 It is known that the distribution of a Gaussian process is determined by its mean and covariance structure. Indeed, the distribution of a process is determined by all joint distributions and the density of a multi-dimensional Gaussian distribution is explicitly given through its mean and covariance matrix. Thus, the two conditions in Definition 1.3.1 determine a unique Gaussian process.

Theorem 1.3.2 $\{B_{1/2}(t)\}$ *is Brownian motion up to a multiplicative constant.*

Proof. Equation (1.3.1) with $H = 1/2$ is the same as in Theorem 1.2.2, and it determines the covariance structure of Brownian motion as mentioned in Remark 1.3.1. $\square$

For the formulation of the next result, we need the notion of a Wiener integral. See Section 3.5 for a definition in a general setting.

Theorem 1.3.3 *A fractional Brownian motion* $\{B_H(t), t \geq 0\}$ *is H-ss with stationary increments. When* $0 < H < 1$*, it has a stochastic integral representation*

$$C_H \left\{ \int_{-\infty}^0 \left((t-u)^{H-1/2} - (-u)^{H-1/2} \right) dB(u) + \int_0^t (t-u)^{H-1/2} dB(u) \right\},$$

(1.3.2)

where

$$C_H = \dot{E} \left[B_H(1)^2 \right]^{1/2} \left\{ \int_{-\infty}^0 \left((t-u)^{H-1/2} - (-u)^{H-1/2} \right)^2 du + \frac{1}{2H} \right\}^{-1/2}.$$

If $H = 1$*, then* $B_1(t) = tB_1(1)$ *almost surely. Fractional Brownian motion is unique in the sense that the class of all fractional Brownian motions coincides with that of all Gaussian selfsimilar processes with stationary increments.* $\{B_H(t)\}$ *has independent increments if and only if* $H = 1/2$*.*

Proof.

    (i) Selfsimilarity. We have that

$$E[B_H(at)B_H(as)] = \frac{1}{2} \left\{ (at)^{2H} + (as)^{2H} - (a|t-s|)^{2H} \right\} E\left[B_H(1)^2 \right]$$

$$= a^{2H} E[B_H(t)B_H(s)]$$

$$= E\left[\left(a^H B_H(t) \right) \left(a^H B_H(s) \right) \right].$$

    Since all processes here are mean zero Gaussian, this equality in covariance implies that $\{B_H(at)\} \stackrel{d}{=} \{a^H B_H(t)\}$.

    (ii) Stationary increments. Again, it is enough to consider only covariances. We have

$$E[(B_H(t+h) - B_H(h))(B_H(s+h) - B_H(h))]$$

$$= E[B_H(t+h)B_H(s+h)] - E[B_H(t+h)B_H(h)]$$

$$- E[B_H(s+h)B_H(h)] + E\left[B_H(h)^2 \right]$$

$$= \frac{1}{2} \left\{ \left((t+h)^{2H} + (s+h)^{2H} - |t-s|^{2H} \right) \right.$$

$$\left. - \left((t+h)^{2H} + h^{2H} - t^{2H} \right) \right.$$

$$-\left((s+h)^{2H}+h^{2H}-s^{2H}\right)+2h^{2H}\Big\}E\left[B_H(1)^2\right]$$

$$=\frac{1}{2}\left(t^{2H}+s^{2H}-|t-s|^{2H}\right)E\left[B_H(1)^2\right]$$

$$=E[B_H(t)B_H(s)],$$

concluding that

$$\{B_H(t+h)-B_H(h)\}\overset{d}{=}\{B_H(t)\}.$$

(iii) For $0<H<1$, the Wiener integral in (1.3.2) is well defined and a mean zero Gaussian random variable. Denote the integral in (1.3.2) by $X(t)$. We then have by Theorem 3.5.1 that

$$E\left[X(t)^2\right]$$

$$=C_H^2\left[\int_{-\infty}^0\left((t-u)^{H-1/2}-(-u)^{H-1/2}\right)^2du+\int_0^t(t-u)^{2H-1}du\right]$$

$$=E\left[B_H(1)^2\right]t^{2H}.$$

Moreover,

$$E\left[(X(t+h)-X(h))^2\right]$$

$$=C_H^2E\left[\left(\int_{-\infty}^h\left((t+h-u)^{H-1/2}-(h-u)^{H-1/2}\right)dB(u)\right.\right.$$

$$\left.\left.+\int_h^{t+h}(t+h-u)^{H-1/2}dB(u)\right)^2\right]$$

$$=C_H^2\Big\{\int_{-\infty}^h\left((t+h-u)^{H-1/2}-(h-u)^{H-1/2}\right)^2du$$

$$+\int_h^{t+h}(t+h-u)^{2H-1}du\Big\}$$

$$=C_H^2\Big\{\int_{-\infty}^0\left((t-u)^{H-1/2}-(-u)^{H-1/2}\right)^2du+\int_0^t(t-u)^{2H-1}du\Big\}$$

$$=E\left[B_H(1)^2\right]t^{2H}.$$

Hence,

$$E[X(t)X(s)] = \frac{1}{2}\left\{E\left[X(t)^2\right] + E\left[X(s)^2\right] - E\left[(X(t) - X(s))^2\right]\right\}$$

$$= \frac{1}{2}\left\{t^{2H} + s^{2H} - |t - s|^{2H}\right\}E\left[B_H(1)^2\right].$$

Therefore, $\{X(t)\}$ for $0 < H < 1$ is fractional Brownian motion.

(iv) For the case $H = 1$, first note that because of (1.3.1), $E[B_1(t)B_1(s)] = ts\, E[B_1(1)^2]$. Then,

$$E\left[(B_1(t) - tB_1(1))^2\right] = E\left[B_1(t)^2\right] - 2tE[B_1(t)\,B_1(1)] + t^2E\left[B_1(1)^2\right]$$

$$= (t^2 - 2t^2 + t^2)E\left[B_1(1)^2\right] = 0,$$

so that $B_1(t) = tB_1(1)$ almost surely.

(v) For the uniqueness, first note that once $\{X(t)\}$ is H-ss and has stationary increments, then by Theorem 1.3.1 above, it has the same covariance structure as in (1.3.1). Since $\{X(t)\}$ is mean zero Gaussian, it is the same as $\{B_H(t)\}$ in law.

(vi) If $H = 1/2$, then by Theorem 1.3.2, the process is Brownian motion. If $\{B_H(t)\}$ has independent increments, then for $0 < s < t$,

$$E[B_H(s)(B_H(t) - B_H(s))] = \frac{1}{2}\left\{t^{2H} + s^{2H} - (t - s)^{2H} - 2s^{2H}\right\}E\left[B_H(1)^2\right]$$

$$= \frac{1}{2}\left\{t^{2H} - s^{2H} - (t - s)^{2H}\right\}E\left[B_H(1)^2\right] = 0.$$

The latter however only holds for $H = 1/2$. □

Remark 1.3.2 Fractional Brownian motion is defined through (1.3.2) in [ManVNe68].

The integral representation of fractional Brownian motion in (1.3.2) is popular, but there is another useful representation through a Wiener integral over a finite interval.

Theorem 1.3.4 [NorValVir99, DecUst99] *When* $0 < H < 1$,

$$B_H(t) \overset{d}{=} C\int_0^t K(t, u)dB(u),$$

where

$$K(t, u) = \left\{\left(\frac{t}{u}\right)^{H-1/2}(t - u)^{H-1/2} - \left(H - \frac{1}{2}\right)u^{1/2-H}\int_u^t x^{H-3/2}(x - u)^{H-1/2}dx\right\}$$

and C is a normalizing constant. For 1/2 < H < 1, a slightly simpler expression for K(t, u) is possible:

$$K(t, u) = \left(H - \frac{1}{2}\right)u^{1/2-H}\int_u^t x^{H-1/2}(x - u)^{H-3/2}dx.$$

1.4 STABLE LÉVY PROCESSES

Definition 1.4.1 *An $\mathbb{R}^d$-valued stochastic process $\{X(t), t \geq 0\}$ is called a Lévy process if*

(a) *$X(0) = 0$ almost surely,*

(b) *it is stochastically continuous at any $t \geq 0$,*

(c) *it has independent and stationary increments, and*

(d) *its sample paths are right-continuous and have left limits almost surely.*

Remark 1.4.1 Excellent references on Lévy processes are [Ber96] and [Sat99]. For an edited volume on the topic with numerous examples, see [BarMikRes01].

Definition 1.4.2 *A probability measure μ on $\mathbb{R}^d$ is called "strictly stable", if for any $a > 0$, there exists $b > 0$ such that $\hat{\mu}(\theta)^a = \hat{\mu}(b\theta)$, $\forall \theta \in \mathbb{R}^d$. In the following, we call such a μ just "stable". If μ is symmetric, it is called "symmetric stable".*

Each stable distribution has a unique index as follows.

Theorem 1.4.1 [SamTaq94] *If μ on $\mathbb{R}^d$ is stable, there exists a unique $\alpha \in (0, 2]$ such that $b = a^{1/\alpha}$. Such a μ is referred to as α-stable. When $\alpha = 2$, μ is a mean zero Gaussian probability measure.*

In terms of independent and identically distributed random variables $X, X_1, X_2, \ldots$ with probability distribution μ, strict stability means that for some $\alpha \in (0, 2]$ $X_1 + \cdots + X_n \overset{d}{\sim} n^{1/\alpha}X$ for all n.

Non-Gaussian stable distributions are sometimes called Lévy distributions by physicists (see [Tsa97]). The special case $\alpha = 1$ is called Cauchy distribution (or Lorentz distribution by physicists). A significant difference between Gaussian distributions and non-Gaussian stable ones like the Cauchy distributions is that the latter have heavy tails, i.e. their variances are infinite. Such models were for a long time not accepted by physicists. More recently, the importance of modeling stochastic phenomena with heavy tailed processes is dramatically increasing in many fields. One important such heavy tail property is the following.

Property 1.4.1 *If Z_α is an $\mathbb{R}^d$-valued random variable with α-stable distribution, $0 < \alpha < 2$, then for any $\gamma \in (0, \alpha)$, $E[|Z_\alpha|^\gamma] < \infty$, but $E[|Z_\alpha|^\alpha] = \infty$.*

Proof. See [SamTaq94], for instance. □

Theorem 1.4.2 *Suppose $\{X(t), t \geq 0\}$ is a Lévy process. Then $\mathcal{L}(X(1))$ is stable if and only if $\{X(t)\}$ is selfsimilar. The index α of stability and the exponent H of selfsimilarity satisfy $\alpha = 1/H$.*

We denote an α-stable Lévy process by $\{Z_\alpha(t), t \geq 0\}$.

Proof. Denote $\mu_t = \mathcal{L}(X(t))$ and $\mu = \mu_1$. Since $\{X(t)\}$ is a Lévy process, for each $t \geq 0$, the characteristic function $\hat{\mu}_t$ satisfies $\hat{\mu}_t(\theta) = \hat{\mu}(\theta)^t$. Indeed, for any n and m

$$X\left(\frac{m}{n}\right) = \left\{X\left(\frac{m}{n}\right) - X\left(\frac{m-1}{n}\right)\right\} + \cdots + \left\{X\left(\frac{1}{n}\right) - X(0)\right\}, \quad (1.4.1)$$

where $X(k/n) - X((k-1)/n)$, $k = 1, \ldots, m$, are independent and identically distributed. It follows from (1.4.1) that $\hat{\mu}_{m/n}(\theta) = \hat{\mu}_{1/n}(\theta)^m$ and in particular that $\hat{\mu}_{1/n}(\theta) = \hat{\mu}(\theta)^{1/n}$. Thus

$$\hat{\mu}_{m/n}(\theta) = \hat{\mu}_{1/n}(\theta)^m = \hat{\mu}(\theta)^{m/n}.$$

This, with the stochastic continuity of $\{X(t)\}$, implies that $\hat{\mu}_t(\theta) = \hat{\mu}(\theta)^t$ for any $t \geq 0$.

We now prove the "if" part of the theorem. By selfsimilarity, for some $H > 0$, $X(a) \stackrel{d}{\sim} a^H X(1)$, $\forall a > 0$, hence $\hat{\mu}(\theta)^a = \hat{\mu}(a^H \theta)$, $\forall a > 0$, $\forall \theta \in \mathbb{R}^d$, implying that μ is stable with $\alpha = 1/H$, necesarily $H \geq \frac{1}{2}$.

For the "only if" part, suppose μ is α-stable, and $0 < \alpha \leq 2$. Since $\{X(t)\}$ has independent and stationary increments, it is enough to show that for any $a > 0$,

$$X(at) \stackrel{d}{\sim} a^{1/\alpha} X(t).$$

However,

$$E[\exp\{i\langle\theta, X(at)\rangle\}] = \hat{\mu}_{at}(\theta) = \hat{\mu}(\theta)^{at} = \hat{\mu}(a^{1/\alpha}\theta)^t = \hat{\mu}_t(a^{1/\alpha}\theta)$$

$$= E\left[\exp\left\{i\langle\theta, a^{1/\alpha}X(t)\rangle\right\}\right].$$

This completes the proof. □

In the following (except for Chapter 9), we restrict ourselves to the case $d = 1$.

1.5 LAMPERTI TRANSFORMATION

Selfsimilar processes are strongly related to strictly stationary processes through a nonlinear time change, referred to as the Lamperti transformation.

Theorem 1.5.1 [Lam62] *If $\{Y(t), t \in \mathbb{R}\}$ is a strictly stationary process and if for some $H > 0$, we let*

$$X(t) = t^H Y(\log t), \quad \text{for } t > 0; \quad X(0) = 0,$$

then $\{X(t), t \geq 0\}$ is H-ss. Conversely, if $\{X(t),\, t \geq 0\}$ is H-ss and if we let

$$Y(t) = e^{-tH} X(e^t), \quad t \in \mathbb{R},$$

then $\{Y(t), t \in \mathbb{R}\}$ is strictly stationary.

Proof. First assume that $\{Y(t), t \in \mathbb{R}\}$ is strictly stationary. For any $n \in \mathbb{N}$, $c_1, \ldots, c_n \in \mathbb{R}$, $t_1, \ldots, t_n > 0$, $a > 0$, we have that

$$\sum_{j=1}^{n} c_j X(at_j) = \sum_{j=1}^{n} c_j a^H t_j^H Y(\log a + \log t_j)$$

$$\overset{d}{\sim} \sum_{j=1}^{n} c_j a^H t_j^H Y(\log t_j)$$

$$= \sum_{j=1}^{n} c_j a^H X(t_j).$$

Thus $\{X(t)\}$ is H-ss. Conversely, for any $n \in \mathbb{N}$, $c_1, \ldots, c_n \in \mathbb{R}$, $t_1, \ldots, t_n > 0$, $h \in \mathbb{R}$, we have that

$$\sum_{j=1}^{n} c_j Y(t_j + h) = \sum_{j=1}^{n} c_j e^{-t_j H} e^{-hH} X(e^h e^{t_j})$$

$$\overset{d}{\sim} \sum_{j=1}^{n} c_j e^{-t_j H} X(e^{t_j})$$

$$= \sum_{j=1}^{n} c_j Y(t_j).$$

Thus $\{Y(t)\}$ is strictly stationary. $\square$

Example 1.5.1 Let $\{B(t), t \geq 0\}$ be standard Brownian motion, which is $\frac{1}{2}$-ss. Then $\{Y(t)\}$ defined by

$$Y(t) = e^{-t/2}B(e^t), \qquad t \in \mathbb{R},$$

is stationary Gaussian, and its covariance is

$$E[Y(t)Y(s)] = e^{-(t+s)/2}E\left[B(e^t)B(e^s)\right]$$

$$= e^{-(t+s)/2}\min\{e^t, e^s\}$$

$$= e^{-|t-s|/2}. \qquad\qquad (1.5.1)$$

Hence, $\{Y(t), t \in \mathbb{R}\}$ is a stationary Ornstein–Uhlenbeck process. This relationship between Brownian motion and a stationary Ornstein–Uhlenbeck process was studied by Doob [Doo42]. The stationary Ornstein–Uhlenbeck process is characterized as a mean zero Gaussian process with covariance (1.5.1). Through the Lamperti transformation, we get a generalization to the stable case. Indeed, if we apply the Lamperti transformation to stable Lévy processes, then it is natural to call the following process a stable stationary Ornstein–Uhlenbeck process:

$$Y(t) = e^{-t/\alpha}Z_\alpha(e^t).$$

This was done in [AdlCamSam90], and leads to an interesting class of stable Markov processes. For more about the Lamperti transformation, see [BurMaeWer95].

Chapter Two

Some Historical Background

2.1 FUNDAMENTAL LIMIT THEOREM

One reason for the fact that selfsimilar processes are important in probability theory is their connection to limit theorems. This was first observed by Lamperti [Lam62]. In the following, we say that a random variable is *nondegenerate* if it is not constant almost surely. The class of slowly varying functions, defined below, will be needed in the formulation of the next theorem.

Definition 2.1.1 *A positive, measurable function L is called slowly varying if for all $x > 0$,*

$$\lim_{t \to \infty} \frac{L(tx)}{L(t)} = 1.$$

A positive, measurable function f is called regularly varying of index $\alpha \in \mathbb{R}$, if $f(x) = x^{\alpha}L(x)$, where L is slowly varying.

For a summary of the basic results on regularly varying functions, see [BinGolTeu87].

Theorem 2.1.1 (Fundamental limit theorem by Lamperti [Lam62]) *Suppose $\mathcal{L}(X(t))$ is nondegenerate for each $t > 0$.*

  (i) *If there exist a stochastic process $\{Y(t), t \geq 0\}$ and real numbers $\{A(\lambda), \lambda \geq 0\}$ with $A(\lambda) > 0$, $\lim_{\lambda \to \infty} A(\lambda) = \infty$ such that as $\lambda \to \infty$,*

$$\frac{1}{A(\lambda)} Y(\lambda t) \stackrel{d}{\Rightarrow} X(t), \tag{2.1.1}$$

    then for some $H > 0$, $\{X(t), t \geq 0\}$ is H-ss.

  (ii) *$A(\lambda)$ in (i) is of the form $A(\lambda) = \lambda^{H}L(\lambda)$, L being a slowly varying function.*

(iii) If $\{X(t), t \geq 0\}$ is H-ss, $H > 0$, then there exist $\{Y(t), t \geq 0\}$ and $\{A(\lambda), \lambda \geq 0\}$ satisfying (2.1.1).

Part (iii) is trivial by taking $\{Y(t)\} = \{X(t)\}$ and $A(\lambda) = \lambda^H$. We present the proofs of parts (i) and (ii). We first state a lemma without proof.

Lemma 2.1.1 [Lam62] *Suppose for distribution functions $\{G_n, n \geq 1\}$, nondegenerate F_1 and F_2, and for real numbers $a_n > 0$, $\alpha_n > 0$, b_n and β_n,*

$$G_n(a_n x + b_n) \longrightarrow F_1(x), \qquad G_n(\alpha_n x + \beta_n) \longrightarrow F_2(x),$$

where the limits are taken at continuity points of the limit distributions, then the following limit exists:

$$\lim_{n \to \infty} \frac{a_n}{\alpha_n} \in (0, \infty).$$

Proof of Theorem 2.1.1. By condition (2.1.1), as $\lambda \to \infty$,

$$P\{Y(\lambda) \leq A(\lambda)x\} \longrightarrow P\{X(1) \leq x\}$$

and

$$P\{Y(\lambda) \leq A(\lambda t)x\} \longrightarrow P\{X(1/t) \leq x\},$$

again for x a continuity point of the limit distributions. By the assumption that $\mathcal{L}(X(t))$ is nondegenerate and the above lemma, we have that, for each $t > 0$,

$$\lim_{\lambda \to \infty} \frac{A(\lambda t)}{A(\lambda)} \in (0, \infty).$$

Then (see, e.g. [BinGolTeu87]), there exist $H \geq 0$ and a slowly varying function $L(\cdot)$ satisfying

$$A(\lambda) = \lambda^H L(\lambda). \tag{2.1.2}$$

Again by (2.1.1), for any $a > 0$, for any $t_1, t_2, \ldots, t_k \geq 0$, and for continuity points $(x_1, \ldots, x_k)$ of the distributions of $(X(t_j), 1 \leq j \leq k)$ and $(X(at_j), 1 \leq j \leq k)$, we have that

$$\lim_{\lambda \to \infty} P\left\{ \frac{1}{A(\lambda)} Y(\lambda t_j) \leq x_j, 1 \leq j \leq k \right\} = P\left\{ X(t_j) \leq x_j, 1 \leq j \leq k \right\}, \tag{2.1.3}$$

$$\lim_{\lambda \to \infty} P\left\{ \frac{1}{A(\lambda)} Y(a\lambda t_j) \leq x_j, 1 \leq j \leq k \right\} = P\left\{ X(at_j) \leq x_j, 1 \leq j \leq k \right\}. \tag{2.1.4}$$

By (2.1.2), the left-hand side of (2.1.4) is

$$\lim_{\lambda \to \infty} P\left\{ \frac{a^H}{A(a\lambda)} Y(a\lambda t_j) \leq \frac{x_j}{h(\lambda)}, 1 \leq j \leq k \right\}$$

for some $h(\lambda) \to 1$. This is equal to

$$P\left\{a^H X(t_j) \le x_j, 1 \le j \le k\right\}$$

by (2.1.3) and the continuous convergence of distribution functions. Since this is equal to the right-hand side of (2.1.4), $\{X(t)\}$ is H-ss.

Under our assumptions, it follows that $H > 0$. Indeed, $A(\lambda) \to \infty$ implies $X(0) = 0$ almost surely. Thus, if $H = 0$, by Theorem 1.1.2, $X(t) = X(0) = 0$ almost surely for all $t > 0$, which contradicts the assumption on $X(t)$ being nondegenerate. □

2.2 FIXED POINTS OF RENORMALIZATION GROUPS

Stable distributions mentioned in Definition 1.4.2 are fixed points of a renormalization group transformation. This fact seems to have been first noted in [Jon75]. Let $R_N, N \ge 1$, be the transformation of a characteristic function $\hat{\mu}(\theta)$ defined by

$$R_N(\hat{\mu})(\theta) = \hat{\mu}(N^{-1/\alpha}\theta)^N.$$

If $\hat{\mu}_\alpha(\theta)$ is the characteristic function of a strictly α-stable random variable, then by Definition 1.4.2 and Theorem 1.4.1,

$$R_N(\hat{\mu}_\alpha) = \hat{\mu}_\alpha.$$

The following result is due to Sinai [Sin76]. Let $H > 0$ and $Y = \{Y_j, j = 0, 1, 2, ...\}$ be a sequence of random variables, and for each $N \ge 1$, define the transformation

$$T_N : Y \to T_N Y = \left\{(T_N Y)_j, j = 0, 1, 2, ...\right\},$$

where

$$(T_N Y)_j = \frac{1}{N^H} \sum_{i=jN}^{(j+1)N-1} Y_i, \qquad j = 0, 1, 2, ...$$

Because $T_N T_M = T_{NM}$, the sequence of transformations $\{T_N, N \ge 1\}$ forms a multiplicative semi-group. It is called the *renormalization group* of index H. Suppose $Y = \{Y_j, j = 0, 1, 2, ...\}$ is a stationary sequence.

Definition 2.2.1 $\mathcal{L}(Y_1)$ is called an *H-selfsimilar distribution* if the stationary sequence $Y = \{Y_j, j = 0, 1, 2, ...\}$ is a fixed point of the renormalization group $\{T_N, N \ge 1\}$ with index H, namely for all $N \ge 1$,

$$\left\{(T_N Y)_j, j = 0, 1, 2, ...\right\} \stackrel{d}{=} \left\{Y_j, j = 0, 1, 2, ...\right\}.$$

Since fractional Brownian motion $\{B_H(t), t \ge 0\}$ has stationary increments, the random variables

$$Y_j = B_H(j + 1) - B_H(j), \qquad j = 0, 1, 2, \ldots$$

form a stationary sequence. The sequence $\{Y_j, j = 1, 2, \ldots\}$ is called *fractional Gaussian noise*.

Theorem 2.2.1 *Within the class of stationary sequences, fractional Gaussian noise is the only Gaussian fixed point of the renormalization group $\{T_N, N \geq 1\}$.*

Proof. For any $\theta_1, \ldots, \theta_k$, $k \geq 0$ and $N \geq 1$,

$$\sum_{j=0}^{k} \theta_j (T_N Y)_j = \sum_{j=0}^{k} \theta_j \frac{1}{N^H} \sum_{i=jN}^{(j+1)N-1} Y_i$$

$$= \sum_{j=0}^{k} \theta_j \frac{1}{N^H} \{B_H((j+1)N) - B_H(jN)\}$$

$$\overset{d}{\sim} \sum_{j=0}^{k} \theta_j \{B_H(j+1) - B_H(j)\}$$

$$= \sum_{j=0}^{k} \theta_j \, Y_j,$$

and thus fractional Gaussian noise is a fixed point of $\{T_N, N \geq 1\}$. Since fractional Brownian motion is the unique Gaussian H-selfsimilar process with stationary increments (see Theorem 1.3.3), fractional Gaussian noise is the unique fixed point. □

Remark 2.2.1 In general, suppose $\{X(t), t \geq 0\}$ is H-ss with stationary increments. (Recall that $X(0) = 0$ almost surely (Property 1.1.1).) Then the increment process

$$Y_j = X(j+1) - X(j), \qquad j = 0, 1, 2, \ldots$$

is a fixed point of the renormalization group transformation $\{T_N, N \geq 1\}$; indeed, the proof of Theorem 2.2.1 also works in this general case.

2.3 LIMIT THEOREMS (I)

Let $X_1, X_2, \ldots$ be a sequence of independent and identically distributed real-valued random variables with $E[X_j] = 0$ and $E[X_j^2] = 1$. Then, as is well known, the central limit theorem holds, namely

$$\frac{1}{n^{1/2}} \sum_{j=1}^{n} X_j \xrightarrow{d} Z, \tag{2.3.1}$$

where Z is a standard normal random variable. Historically, the next question was how we can relax the assumption on independence of $\{X_j\}$ while keeping the validity of the central limit theorem (2.3.1). Rosenblatt [Ros56] introduced a mixing condition which is a kind of weak dependence condition for stationary sequences of random variables. Numerous extensions to other mixing conditions have since been introduced. An important problem addressed by Rosenblatt was as follows: suppose that a stationary sequence has a stronger dependence violating the validity of the central limit theorem, then what type of limiting distributions are expected to appear. He answered this question in [Ros61] laying the foundation of the so-called noncentral limit theorems.

Theorem 2.3.1 *Let $\{\xi_j\}$ be a stationary Gaussian sequence such that $E[\xi_0] = 0$, $E[\xi_0^2] = 1$ and $E[\xi_0\xi_k] \sim k^{H-1}L(k)$ as $k \to \infty$ for some $H \in (1/2, 1)$ and some slowly varying function L. Define another stationary sequence $\{X_j\}$ by*

$$X_j = \xi_j^2 - 1.$$

Then

$$\frac{1}{n^H L(n)} \sum_{j=1}^{n} X_j \xrightarrow{d} Z, \tag{2.3.2}$$

where Z is a non-Gaussian random variable given by

$$E\left[e^{i\theta Z}\right] = \exp\left\{\sum_{k=2}^{\infty} \frac{(2i\theta)^k}{2k} \int_{x \in [0,1]^k} |x_1 - x_k|^{-2(1-H)} \right.$$

$$\left. \prod_{j=2}^{k} |x_j - x_{j-1}|^{-2(1-H)} dx \right\}, \qquad \theta \in \mathbb{R}.$$

As we will see in Section 3.4, Taqqu [Taq75] considered the "process version" of (2.3.2) and obtained a limiting process of

$$X_n(t) := \frac{1}{n^H L(n)} \sum_{j=1}^{[nt]} X_j.$$

This limiting process is, of course, H-ss by Theorem 2.1.1, and the first example of non-Gaussian selfsimilar processes having strongly dependent increment structure. It is referred to as the *Rosenblatt process*.

Chapter Three

Selfsimilar Processes with Stationary Increments

Stable Lévy processes (including Brownian motion) are the only selfsimilar processes with independent and stationary increments. Consequently, one is interested in selfsimilar processes with just stationary increments or just independent increments. In this chapter, we discuss selfsimilar processes with stationary increments. Selfsimilar processes with independent increments are discussed in Chapter 5.

When $\{X(t), t \geq 0\}$ is H-selfsimilar with stationary increments, we call it H-ss, si, for short.

3.1 SIMPLE PROPERTIES

The following results give some basic formulas and estimates on moments and the exponent of selfsimilarity.

Theorem 3.1.1 *Suppose that $\{X(t)\}$ is H-ss, si, $H > 0$ and that $X(t)$ is nondegenerate for each $t > 0$.*

(i) *[Mae86] If $E[|X(1)|^\gamma] < \infty$ for some $0 < \gamma < 1$, then $H < 1/\gamma$.*

(ii) *If $E[|X(1)|] < \infty$, then $H \leq 1$.*

(iii) *[Kon84] If $E[|X(1)|] < \infty$ and $0 < H < 1$, then $E[X(t)] = 0$ for all $t \geq 0$.*

(iv) *If $E[|X(1)|] < \infty$ and $H = 1$, then $X(t) = tX(1)$ almost surely.*

Before proving Theorem 3.1.1, we will provide some remarks.

Remark 3.1.1 The inequality in Theorem 3.1.1 (i) is best possible. As an example, consider an α-stable Lévy process with $\alpha < 1$ (Section 1.4), for which $H = 1/\alpha$. On the other hand, by Property 1.4.1, $E[|Z_\alpha(1)|^\gamma] < \infty$ for any $0 < \gamma < \alpha$ and $E[|Z_\alpha(1)|^\alpha] = \infty$. Hence H cannot be greater than or equal to $1/\gamma$. What will happen if we drop the assumption of stationary increments in the above. For any $H > 0$ and any random variable ξ,

$$X(t) := t^H \xi \tag{3.1.1}$$

is H-ss. Hence we cannot say anything about $X(1)$ just from H-ss. If $H \neq 1$, (3.1.1) does not have stationary increments, but if $H = 1$, it is 1-ss, si. Thus, when $H = 1$, 1-ss, si is not sufficient to get some information about $X(1)$ (see for instance (iii) above). As follows from (iv), when $E[|X(1)|] < \infty$, a 1-ss, si process is always of the form (3.1.1).

Proof of Theorem 3.1.1. (i) Note that if $a_j \geq 0$ for all $1 \leq j \leq N$, $a_m > 0$, $a_n > 0$ for some $1 \leq m \neq n \leq N$, and if $0 < \gamma < 1$, then

$$\left(\sum_{j=1}^{N} a_j \right)^{\gamma} < \sum_{j=1}^{N} a_j^{\gamma}. \tag{3.1.2}$$

We now see from the assumption that X is nondegenerate that

$$\varepsilon := P\{X(1) \neq 0\} > 0.$$

Let N be an integer satisfying $N > 1/\varepsilon$. Recall that $X(0) = 0$ almost surely (Property 1.1.1). By si, we see that

$$P\{X(j) - X(j-1) \neq 0\} = \varepsilon > 0, \qquad 1 \leq j \leq N,$$

from which there exist m and n with $1 \leq m < n \leq N$ such that

$$A := \{\omega \in \Omega, X(m, \omega) - X(m-1, \omega) \neq 0 \text{ and } X(n, \omega) - X(n-1, \omega) \neq 0\}$$

has a strictly positive probability. Hence it follows from (3.1.2) that for any $\omega \in A$,

$$|X(N, \omega)|^{\gamma} \leq \left(\sum_{j=1}^{N} |X(j, \omega) - X(j-1, \omega)| \right)^{\gamma}$$

$$< \sum_{j=1}^{N} |X(j, \omega) - X(j-1, \omega)|^{\gamma}. \tag{3.1.3}$$

Since the relation (3.1.3) also holds with $\leq$ for any $\omega \in A^c$ and $P(A) > 0$, it follows that

$$E[|X(N)|^{\gamma}] < \sum_{j=1}^{N} E[|X(j) - X(j-1)|^{\gamma}].$$

Thus by H-ss, si,

$$N^{H\gamma} E[|X(1)|^{\gamma}] < N E[|X(1)|^{\gamma}],$$

implying $H < 1/\gamma$.

(ii) Since $E[|X(1)|] < \infty$, $E[|X(1)|^{\gamma}] < \infty$ for any $\gamma < 1$. Thus, by (i), $H < 1/\gamma$ for any $\gamma < 1$, meaning $H \leq 1$.

(iii) By H-ss, si, it follows that $E[X(t)] = E[X(2t) - X(t)] = (2^H - 1)E[X(t)]$, and so $(2^H - 2)E[X(t)] = 0$. Since $0 < H < 1$, we have that $E[X(t)] = 0$.

(iv) The general proof under the assumption $E[|X(t)|] < \infty$ is omitted (see [Ver85]). A simple proof when $E[X(1)^2] < \infty$ was given in part (iv) of the proof of Theorem 1.3.3 for fractional Brownian motion. Exactly the same reasoning applies here. $\square$

Recall that an α-stable Lévy process $\{Z_\alpha(t)\}$ for $0 < \alpha < 2$ satisfies $E[|Z_\alpha(1)|^\alpha] = \infty$. In other words, if $\{X(t)\}$ is H-ss, si and $\{X(t)\} \stackrel{d}{=} \{Z_\alpha(t)\}$, then $E[|X(1)|^{1/H}] = \infty$. This is also true for any H-ss, si process with $H > 1$.

Theorem 3.1.2 [Mae86] *Let $\{X(t)\}$ be H-ss, si and $H > 1$. Then $E[|X(1)|^{1/H}] = \infty$.*

Proof. Suppose $E[|X(1)|^{1/H}] < \infty$. Since $\gamma := 1/H < 1$, by Theorem 3.1.1 (i), we have that $H < 1/\gamma = H$. This is a contradiction. $\square$

3.2 LONG-RANGE DEPENDENCE (I)

Let $\{X(t), t \geq 0\}$ be H-ss, si, $0 < H < 1$, and nondegenerate for each $t > 0$ with $E[X(1)^2] < \infty$, and define

$$\xi(n) = X(n + 1) - X(n), \qquad n = 0, 1, 2, \ldots,$$

$$r(n) = E[\xi(0)\xi(n)], \qquad n = 0, 1, 2, \ldots.$$

Then

$$r(n) \sim H(2H - 1)n^{2H-2}E[X(1)^2] \quad \text{as } n \to \infty, \text{ if } H \neq \frac{1}{2}, \qquad (3.2.1)$$

$$r(n) = 0, \qquad n \geq 1, \qquad \text{if } H = \frac{1}{2}.$$

This can be shown as follows. Noticing that $X(0) = 0$ almost surely (Property 1.1.1) and using Theorem 1.3.1, we have, for $n \geq 1$,

$$r(n) = E[\xi(0)\xi(n)] = E[X(1)\{X(n + 1) - X(n)\}]$$

$$= E[X(1)X(n + 1)] - E[X(1)X(n)]$$

$$= \frac{1}{2}\{(n + 1)^{2H} - 2n^{2H} + (n - 1)^{2H}\}E[X(1)^2],$$

which implies the conclusion. Hence

(a) if $0 < H < 1/2$, $\sum_{n=0}^{\infty} |r(n)| < \infty$,

(b) if $H = 1/2$, $\{\xi(n)\}$ is uncorrelated,

(c) if $1/2 < H < 1$, $\sum_{n=0}^{\infty} |r(n)| = \infty$.

Actually, if $0 < H < 1/2$, $r(n) < 0$ for $n \geq 1$ (negative correlation), if $1/2 < H < 1$, $r(n) > 0$ for $n \geq 1$ (positive correlation). The property $\sum |r(n)| = \infty$ is often referred to as long-range dependence and is especially of interest in statistics. (See [Ber94] and [Cox84]).

3.3 SELFSIMILAR PROCESSES WITH FINITE VARIANCES

In this section, we explain when stochastic processes with finite variances, which can be represented by multiple Wiener–Itô integrals, are ss, si processes. We follow [Dob79] and [Maj81b].

For a spectral measure G on $\mathbb{R}$, define a complex-valued random spectral measure Z_G on the Borel σ-algebra $\mathfrak{B}(\mathbb{R})$ satisfying the following conditions:

(i) $Z_G(A)$ is a complex-valued Gaussian random variable,

(ii) $E[Z_G(A)] = 0$,

(iii) $E[Z_G(A)\overline{Z_G(B)}] = G(A \cap B)$,

(iv) $Z_G(\bigcup_{j=1}^{n} A_j) = \sum_{j=1}^{n} Z_G(A_j)$ for mutually disjoint $A_1, ..., A_n$,

(v) $Z_G(A) = \overline{Z_G(-A)}$.

By H_G^k we represent the class of complex-valued functions f of k variables satisfying the following properties:

(i) $f(-x_1, ..., -x_k) = \overline{f(x_1, ..., x_k)}$,

(ii) $\int_{\mathbb{R}^k} |f(x_1, ..., x_k)|^2 G(dx_1) \cdots G(dx_k) < \infty$,

(iii) $f(x_{i_1}, ..., x_{i_k}) = f(x_1, ..., x_k)$, where $\{i_1, ..., i_k\} = \{1, ..., k\}$.

Then we can define for any $f \in H_G^k$, the *multiple Wiener–Itô integral*

$$\int_{\mathbb{R}^k}'' f(x_1, ..., x_k) Z_G(dx_1) \cdots Z_G(dx_k). \tag{3.3.1}$$

Here $\int_{\mathbb{R}^k}''$ is the integral over $\mathbb{R}^k$ except the hyperplanes $x_i = \pm x_j$, $i \neq j$.

For later use, we state here the *change of variables formula* for multiple Wiener–Itô integrals (see [Maj81b] for details). Let G and $\tilde{G}$ be nonatomic spectral measures such that G is absolutely continuous with respect to $\tilde{G}$ and suppose that $g(x)$ is a complex-valued function satisfying $g(x) = \overline{g(-x)}$ and $|g(x)|^2 = dG(x)/d\tilde{G}(x)$. For $f \in H_G^k$, put

$$\tilde{f}(x_1, ..., x_k) = f(x_1, ..., x_k) g(x_1) \cdots g(x_k).$$

Then $\tilde{f} \in H_{\tilde{G}}^k$ and

$$\int_{\mathbb{R}^k}'' f(x_1, \ldots, x_k) Z_G(dx_1) \cdots Z_G(dx_k)$$

$$\overset{d}{\sim} \int_{\mathbb{R}^k}'' \tilde{f}(x_1, \ldots, x_k) Z_{\tilde{G}}(dx_1) \cdots Z_{\tilde{G}}(dx_k), \tag{3.3.2}$$

where Z_G and $Z_{\tilde{G}}$ are random spectral measures corresponding to G and $\tilde{G}$, respectively.

Dobrushin [Dob79] gave a condition for processes having the multiple Wiener–Itô integral representation to be selfsimilar.

Theorem 3.3.1 [Dob79] *For $k > 0$ and $f \in H_G^k$, put*

$$X(t) = \int_{\mathbb{R}^k}'' f(x_1, \ldots, x_k) \phi_t(x_1 + \cdots + x_k) Z_G(dx_1) \cdots Z_G(dx_k), \tag{3.3.3}$$

where $\phi_t(x) = (e^{itx} - 1)/ix$. If, for some $p > 0$ and $q \in \mathbb{R}$,

(a) $f(cx_1, \ldots, cx_k) = c^p f(x_1, \ldots, x_k)$,

(b) $G(cA) = c^q G(A)$, $A \in \mathfrak{B}(\mathbb{R})$,

then $\{X(t)\}$ is H-ss, si, where $H = 1 - p - kq/2$.

Since $E[X(t)^2] < \infty$, it follows from Theorem 3.1.1 (ii) and (iv) that p and q must satisfy $0 < p + kq/2 < 1$. The proof of this theorem is easy if we use the change of variables formula (3.3.2).

Among the selfsimilar processes characterized by Theorem 3.3.1, we consider those special cases which will appear as limits in the noncentral limit theorem in the next section. Suppose $f \equiv 1$ in (3.3.3). Namely consider

$$X(t) = \int_{\mathbb{R}^k}'' \phi_t(x_1 + \cdots + x_k) Z_G(dx_1) \cdots Z_G(dx_k). \tag{3.3.4}$$

This process is called the *Hermite process* and, in particular, the *Rosenblatt process* when $k = 2$ as mentioned in Section 2.3. When $k = 1$, it is Gaussian and when $k \geq 2$, non-Gaussian. Since condition (a) in Theorem 3.3.1 is trivially satisfied with $p = 0$, $X(t)$ is H-ss with $H = 1 - kq/2$ if the process exists. (Actually, for general k, the definition is only meaningful when $0 < q < 1/k$, and so H satisfies $1/2 < H < 1$.)

$\{X(t)\}$ in (3.3.4) can also be represented by an integral with respect to standard Brownian motion $\{B(t), t \in \mathbb{R}\}$ in the following way:

$$X(t) \overset{d}{=} C_q \int_{\mathbb{R}^k}' \int_0^t \left(\prod_{j=1}^k (s - y_j)^{-(1+q)/2} I[y_j < s] \right) ds \, dB(y_1) \cdots dB(y_k), \tag{3.3.5}$$

where $\int'_{\mathbb{R}^k}$ is the integral over $\mathbb{R}^k$ except the hyperplanes $x_i = x_j$, $i \neq j$, $I[\cdot]$ is the indicator function and C_q is a constant depending only on q. The identity (3.3.5) can be shown by using Parseval's identity for multiple Wiener–Itô integrals [Taq81]; for $h \in L^2(\mathbb{R}^k)$,

$$\int'_{\mathbb{R}^k} h(y_1, ..., y_k) \, dB(y_1) \cdots dB(y_k)$$

$$\overset{d}{\sim} \int''_{\mathbb{R}^k} \hat{h}(x_1, ..., x_k) |x_1|^{(1-q)/2} \cdots |x_k|^{(1-q)/2} Z_G(dx_1) \cdots Z_G(dx_k),$$

where $\hat{h}$ is the Fourier transform of h on $\mathbb{R}^k$. Especially, if

$$h(y_1, ..., y_k) = \int_0^t \prod_{j=1}^{k} (s - y_j)^{-(1+q)/2} I[y_j < s] ds,$$

then

$$\hat{h}(x_1, ..., x_k) = C_q^{-1} \phi_t(x_1 + \cdots + x_k) |x_1|^{(q-1)/2} \cdots |x_k|^{(q-1)/2}.$$

3.4 LIMIT THEOREMS (II)

Extending the idea of Rosenblatt (Theorem 2.3.1), Taqqu [Taq75, Taq79], and independently Dobrushin and Májòr [DobMaj79], proved the following noncentral limit theorem, which has the Hermite processes as limits.

Let $\{\xi_n, n \in \mathbb{Z}\}$ be a sequence of stationary Gaussian random variables with $E[\xi_0] = 0$, $E[\xi_0^2] = 1$, and further assume that the covariances satisfy

$$r(n) = E[\xi_0 \xi_n] \sim n^{-q} L(n), \qquad n \to \infty, \tag{3.4.1}$$

where $0 < q < 1$ and L is a slowly varying function. (In this case, of course, L needs not be positive; see [BinGolTeu87] for this trivial extension of the notion of slow variation.) Let G be the spectral measure of $\{\xi_n\}$ such that $r(n) = \int_{-\pi}^{\pi} e^{inx} G(dx)$.

Lemma 3.4.1 [DobMaj79] *Define* $\{G_n, n = 1, 2, ...\}$ *by*

$$G_n(A) = \frac{n^q}{L(n)} G\left(\frac{A}{n} \cap [-\pi, \pi)\right), \qquad A \in \mathfrak{B}(\mathbb{R}). \tag{3.4.2}$$

Then there exists a locally finite measure G_0 *such that* $G_n \to G_0$ *(vaguely) and for any* $c > 0$, $A \in \mathfrak{B}(\mathbb{R})$, $G_0(cA) = c^q G_0(A)$.

(See, e.g. [EmbKluMik97, p. 563], for a definition of vague convergence.)

Let Z_{G_0} be the random spectral measure corresponding to G_0 and put

$$X_0(t) = \int_{\mathbb{R}^k} \phi_t(x_1 + \cdots + x_k)\, Z_{G_0}(dx_1)\cdots Z_{G_0}(dx_k). \qquad (3.4.3)$$

This is the Hermite process defined in (3.3.4).

Let f be a function satisfying $E[f(\xi_0)] = 0$, $E[f(\xi_0)^2] < \infty$, and expand f by Hermite polynomials as

$$f(x) = \sum_{j=0}^{\infty} c_j H_j(x),$$

where $H_j(x)$ is the Hermite polynomial whose leading coefficient is 1, $c_j = (1/j!)E[f(\xi_0)H_j(\xi_0)]$, and the convergence is taken in the sense of mean square. Define

$$k = \min\{j \mid c_j \neq 0\},$$

this k being referred to as the *Hermite rank* of f. By the assumption $E[f(\xi_0)] = 0$, $c_0 = 0$ so that $k \geq 1$.

Theorem 3.4.1 (Noncentral limit theorem [DobMaj79, Taq79]) *Let k be the Hermite rank of f and $\{\xi_n, n \in \mathbb{Z}\}$ stationary Gaussian random variables introduced at the beginning of this section, and assume that (3.4.1) holds for some q with $0 < q < 1/k$. (We define G_0 in Lemma 3.4.1 by using this q and further $X_0(t)$ by (3.4.3).) If $A_n = n^{1-kq/2}L(n)^{k/2}$, then*

$$X_n(t) := \frac{1}{A_n}\sum_{j=1}^{[nt]} f(\xi_j) \overset{d}{\Rightarrow} c_k X_0(t).$$

Notice that the multiplicity k of the integral of the limiting ss process $\{X_0(t)\}$ is identical to the Hermite rank of f.

Sketch of the proof of Theorem 3.4.1. If we put

$$f(x) = c_k H_k(x) + f_k^*(x),$$

where $f_k^*(x) = \sum_{j=k+1}^{\infty} c_j H_j(x)$, we can easily verify that, under our assumptions,

$$E\left[\left|\frac{1}{A_n}\sum_{j=1}^{[nt]} f_k^*(\xi_j)\right|^2\right] \to 0.$$

For this, the condition $q < 1/k$ plays an essential role. Hence it is enough to show that, when $f(x) = H_k(x)$, $X_n(t)$ converges to $X_0(t)$. Note that

$$\xi_j = \int_{-\pi}^{\pi} e^{ijx} Z_G(dx),$$

where Z_G is the random spectral measure corresponding to the spectral measure G of the stationary sequence $\{\xi_j\}$. On the other hand, it is known that

$$
H_k\left(\int_{-\pi}^{\pi} e^{ijx} Z_G(dx)\right) \overset{d}{\sim} \int_{[-\pi,\pi)^k}'' e^{ij(x_1+\cdots+x_k)} Z_G(dx_1)\cdots Z_G(dx_k)
$$

[Ito51]. From these, we have that

$$
X_n(t) = \frac{1}{A_n}\int_{[-\pi,\pi)^k}'' \frac{e^{i(x_1+\cdots+x_k)}\left(e^{i[nt](x_1+\cdots+x_k)}-1\right)}{e^{i(x_1+\cdots+x_k)}-1} Z_G(dx_1)\cdots Z_G(dx_k),
$$

and by the change of variables $x_j = y_j/n$,

$$
X_n(t) = \frac{1}{n^{1-kq/2}L(n)^{k/2}}\int_{[-n\pi,n\pi)^k}'' \frac{e^{i(y_1+\cdots+y_k)/n}}{e^{i(y_1+\cdots+y_k)/n}-1}
$$

$$
\left(e^{i([nt]/n)(y_1+\cdots+y_k)}-1\right)Z_G\left(\frac{dy_1}{n}\right)\cdots Z_G\left(\frac{dy_k}{n}\right).
$$

If we define G_n as in (3.4.2) of Lemma 3.4.1 and again use the change of variables formula, we can rewrite the above as an integral with respect to the random spectral measure Z_{G_n} corresponding to G_n:

$$
X_n(t) \overset{d}{=} \frac{1}{n}\int_{[-n\pi,n\pi)^k}'' \frac{e^{i(y_1+\cdots+y_k)/n}}{e^{i(y_1+\cdots+y_k)/n}-1}\left(e^{i([nt]/n)(y_1+\cdots+y_k)}-1\right)Z_{G_n}(dy_1)\cdots Z_{G_n}(dy_k)
$$

$$
= \int_{[-n\pi,n\pi)^k}'' f_n(y_1,\ldots,y_k)\phi_t(y_1+\cdots+y_k)Z_{G_n}(dy_1)\cdots Z_{G_n}(dy_k),
$$

where

$$
f_n(y_1,\ldots,y_k) = e^{i(y_1+\cdots+y_k)/n}\,\frac{i(y_1+\cdots+y_k)/n}{e^{i(y_1+\cdots+y_k)/n}-1}\cdot\frac{e^{i([nt]/n)(y_1+\cdots+y_k)}-1}{e^{it(y_1+\cdots+y_k)}-1},
$$

and ϕ_t is defined in (3.3.3). Then $f_n \to 1$ (uniformly on every bounded domain) and $G_n \to G_0$ (vaguely, by Lemma 3.4.1). The limit $X_0(t)$ is obtained by replacing f_n by 1 and G_n by G_0 in the above expression for $X_n(t)$. It remains to justify this interchange of limits; this can be done by a standard Fourier transform argument. $\square$

Remark 3.4.1 In Theorem 3.4.1, the condition $0 < q < 1/k$ is essential for the validity of the noncentral limit theorem. This condition assures that the order of $\mathrm{Var}(\sum_{j=1}^{n} f(\xi_j))$ is greater than n, implying that the random variables $\{f(\xi_j), j = 1, 2, \ldots\}$ are strongly dependent. This is the reason why non-Gaussian limits appear and why the theorem is called the noncentral limit theorem. What will happen if the order of $\mathrm{Var}(\sum_{j=1}^{n} f(\xi_j))$ is n or $nL(n)$, L being slowly varying? This corresponds to the case $q \geq 1/k$, and it is known that the central limit theorem again holds [BreMaj83, GirSur85, Mar76, Mar80].

Another interesting limit theorem is given by Májòr [Maj81a]. Under the same condition as in Theorem 3.4.1, put

$$c_j = \begin{cases} |j|^{\gamma-1}\text{sgn}(j), & j \neq 0, \\ 0, & j = 0, \end{cases}$$

with $kq/2 - 1 < \gamma < kq/2$, $\gamma \neq 0$, and define

$$\zeta_l = \sum_{j=-\infty}^{\infty} c_j H_k(\xi_{l+j}).$$

Here $H_k(\cdot)$ is the kth Hermite polynomial, as before.

Theorem 3.4.2 [Maj81a]   *Under the same assumptions as in Theorem 3.4.1, if we take* $A_n = n^{1+\gamma-kq/2}L(n)^{k/2}$, *then*

$$\frac{1}{A_n} \sum_{l=1}^{[nt]} \zeta_l \overset{d}{\Rightarrow} X(t),$$

where

$$X(t) = C \int_{\mathbb{R}^k}'' \phi_t \left(x_1 + \cdots + x_k\right) i|x_1 + \cdots + x_k|^{-\gamma}$$

$$\text{sgn}(x_1 + \cdots + x_k) \, Z_{G_0}(dx_1)\cdots Z_{G_0}(dx_k),$$

and G_0 is also the same as in Theorem 3.4.1.

This limiting process $\{X(t)\}$ belongs to the class of selfsimilar processes constructed in Theorem 3.3.1, and is H-ss, si, with $H = 1 + \gamma - kq/2 \in (0, 1)$.

Remark 3.4.2   If we take $\gamma = kq/2 - 1/2$, then $H = 1/2$, which is the same as Brownian motion. However, unless $k = 1$, it is non-Gaussian hence not Brownian motion. On the other hand, this process has properties similar to Brownian motion, namely, $E[X(t)\overline{X(s)}] = \min\{t, s\}$ and its disjoint increments are uncorrelated as seen in Section 3.2; they are, however, not independent.

3.5 STABLE PROCESSES

In this section, we discuss some basic facts on stable processes for later use. Most results in the following three sections can also be found in the excellent book on stable processes by Samorodnitsky and Taqqu [SamTaq94]. Recall that $\{X(t)\}$ is a Gaussian process if all joint distributions are Gaussian. Stable processes are a generalization of Gaussian processes. The marginal distribution of a Gaussian process, centered at the origin, is symmetric. So for simpli-

city, we restrict ourselves here to the symmetric case. We have already given
the definition of a symmetric stable random variable (see Definition 1.4.2).

It is known that a real-valued random variable ξ is symmetric α-stable
($S\alpha S$), $0 < \alpha \le 2$, if and only if its characteristic function satisfies

$$E\left[e^{i\theta\xi}\right] = e^{-c|\theta|^\alpha}, \qquad \theta \in \mathbb{R},$$

for some $c > 0$ (scaling parameter). We write $\xi \sim S\alpha S(c)$ when we want to
emphasize the scaling parameter c.

Remark 3.5.1 For $S2S$, i.e. $\alpha = 2$, we have the Gaussian case.

Definition 3.5.1 *A real-valued stochastic process $\{X(t)\}$ is said to be $S\alpha S$ if
any linear combination $\sum_{k=1}^n a_k X(t_k)$, $a_k \in \mathbb{R}$, is $S\alpha S(c)$, where $c > 0$ depends
on $(a_1, ..., a_n, t_1, ..., t_n)$.*

In the symmetric case, stability of all linear combinations implies stability of
all joint distributions; see [SamTaq94, Theorem 2.1.5]. In order to construct
examples of selfsimilar processes with infinite variance, we need the notion of
an integral with respect to a symmetric stable Lévy process $\{Z_\alpha(t), t \in \mathbb{R}\}$,
including the Wiener integral which is an integral with respect to Brownian
motion. Here $\{Z_\alpha(t), t \ge 0\}$ is a Lévy process such that $\mathcal{L}(Z_\alpha(t))$ is $S\alpha S$. We
extend the definition of $\{Z_\alpha(t), t \ge 0\}$ to the whole of $\mathbb{R}$ in the following way.
Let $\{Z_\alpha^{(-)}(t), t \ge 0\}$ be an independent copy of $\{Z_\alpha(t), t \ge 0\}$ and define for
$t < 0$, $Z_\alpha(t) = Z_\alpha^{(-)}(-t)$. Note that $\{Z_\alpha(t)\}$ satisfies

$$E\left[e^{i\theta Z_\alpha(t)}\right] = e^{-c|t||\theta|^\alpha}, \qquad \theta \in \mathbb{R}.$$

Without loss of generality on scaling, we assume that $c = 1$. Therefore, for
each t, $Z_\alpha(t) \sim S\alpha S(|t|)$.

Theorem 3.5.1 *Let $0 < \alpha \le 2$ and $A \subset \mathbb{R}$. If*
$$\int_A |f(x)|^\alpha \, dx < \infty,$$

then a stable integral

$$I(f) := \int_A f(x) \, dZ_\alpha(x)$$

can be defined in the sense of convergence in probability. Further,

$$I(f) \sim S\alpha S\left(\int_A |f(x)|^\alpha \, dx\right),$$

that is, $I(f)$ is $S\alpha S$. When $\alpha = 2$, $E[(\int_A f(x) \, dB(x))^2] = \int_A f(x)^2 dx$.

The proof of this result can be found in [SamTaq94]. One first verifies the theorem for simple functions $f(x) = \sum_{j=1}^{n} c_j 1_{(t_{j-1}, t_j]}(x)$ and then passes suitably to a limit.

Theorem 3.5.2 *If $\{f_t(\cdot), t \geq 0\}$ is a set of measurable functions and if*

$$\int_E |f_t(x)|^{\alpha} dx < \infty, \qquad \text{for each } t \geq 0,$$

then the process $\{X(t), t \geq 0\}$, defined by

$$X(t) = \int_E f_t(x) dZ_{\alpha}(x), \qquad t \geq 0,$$

is a SαS process.

Proof. It suffices to apply Definition 3.5.1 and Theorem 3.5.1 to the integrand $\sum_{k=1}^{n} a_k f_{t_k}(x)$. $\square$

The following result on independence of stable integrals will be used later.

Theorem 3.5.3 [Sch70, Har82], see also Theorem 3.5.3 in [SamTaq94]
Let $0 < \alpha < 2$. Two stable random variables $I(f)$ and $I(g)$ are independent if and only if $f(x)g(x) = 0$ almost everywhere.

Remark 3.5.2 When $\alpha = 2$, Theorem 3.5.3 does not hold. For an easy example take $f(x) = 1$, $x \in [0, 2]$ and $g(x) = -1$, $x \in [0, 1]$, $g(x) = 1$, $x \in (1, 2]$; then $\int_0^2 f(x) \, dB(x) = B(2)$, $\int_0^2 g(x) \, dB(x) = B(2) - 2B(1)$, which are clearly uncorrelated, and hence independent.

3.6 SELFSIMILAR PROCESSES WITH INFINITE VARIANCE

In this section, we discuss selfsimilar symmetric stable processes with stationary increments (abbreviated as H-ss, si, SαS processes). We recall the characterizations for Brownian motion and the stable Lévy process in terms of selfsimilarity, as mentioned in Theorem 1.4.2. Namely, suppose that the process $\{X(t)\}$ is H-ss ($H > 0$), si, ii (independent increments) and SαS, $0 < \alpha \leq 2$. Then $\{X(t)\}$ is a Brownian motion when $\alpha = 2$ and a SαS Lévy process when $0 < \alpha < 2$, and $H = 1/\alpha$. Because of this fact, the processes we will treat are selfsimilar processes with dependent increments.

H-ss, si, SαS processes have two parameters H and α ($H > 0$ and $0 < \alpha \leq 2$). However, there is a restriction for H and α.

Theorem 3.6.1 [KasMaeVer88] *For $H > \max(1, 1/\alpha)$, such an H-ss, si, SαS process does not exist.*

Proof. Let $\{X(t)\}$ be *H*-ss, si, *SαS*. It is known that $E[|X(1)|^\gamma] < \infty$ for any $\gamma \in (0, \alpha)$, but $E[|X(1)|^\alpha] = \infty$ if $\alpha < 2$. Suppose $\alpha > 1$. Then $E[|X(t)|] < \infty$. By Theorem 3.1.1 (ii), $H \leq 1$. Suppose $0 < \alpha \leq 1$. Then by Theorem 3.1.1 (i), $H < 1/\gamma$ for any $0 < \gamma < \alpha$. Thus $H \leq 1/\alpha$, and *H* cannot be greater than $\max(1, 1/\alpha)$. $\square$

A generalization of fractional Brownian motion to the case $0 < \alpha < 2$ leads to *linear fractional stable motion*.

Example 3.6.1 [TaqWol83, Mae83, KasMae88] Let $0 < H < 1$, $0 < \alpha \leq 2$, $a, b \in \mathbb{R}$, $|a| + |b| > 0$. *Linear fractional stable motion* $\{\Delta_{H,\alpha}(a, b; t), t \geq 0\}$ is defined by

$$\Delta_{H,\alpha}(a, b; t) = \int_{-\infty}^{\infty} \left[a \left\{ (t - u)_+^{H-1/\alpha} - (-u)_+^{H-1/\alpha} \right\} \right.$$

$$\left. + b \left\{ (t - u)_-^{H-1/\alpha} - (-u)_-^{H-1/\alpha} \right\} \right] dZ_\alpha(u), \quad (3.6.1)$$

where $x_+ = \max(x, 0)$, $x_- = \max(-x, 0)$ and $0^s = 0$, even for $s \leq 0$. $\Delta_{1/\alpha,\alpha}(1, 0; t)$ is a *SαS* Lévy process.

Theorem 3.6.2 *The linear fractional stable motion $\{\Delta_{H,\alpha}(a, b; t)\}$ is H-ss, si, SαS.*

Proof. Since

$$f_t(u) := a \left\{ (t - u)_+^{H-1/\alpha} - (-u)_+^{H-1/\alpha} \right\} + b \left\{ (t - u)_-^{H-1/\alpha} - (-u)_-^{H-1/\alpha} \right\}$$

is in $L^\alpha(\mathbb{R})$ if $0 < H < 1$ and $0 < \alpha \leq 2$, $\{\Delta_{H,\alpha}(a, b; t)\}$ is a *SαS* process by Theorem 3.5.2. The property of *H*-ss can be proved in the following way. Consider $\Delta_{H,\alpha}(a, b; ct)$ in (3.6.1), change the variable *u* to *cv* and use the selfsimilarity of Z_α, i.e. $Z_\alpha(cv) \stackrel{d}{=} c^{1/\alpha} Z_\alpha(v)$, to obtain

$$\Delta_{H,\alpha}(a, b; ct) \stackrel{d}{=} c^H \Delta_{H,\alpha}(a, b; t).$$

That $\Delta_{H,\alpha}(a, b; \cdot)$ has stationary increments follows from the si property of Z_α, that is, $Z_\alpha(u + h) - Z_\alpha(h) \stackrel{d}{=} Z_\alpha(u)$. $\square$

Suppose $\alpha = 2$. Then for each $H \in (0, 1)$, all linear fractional stable motions $\{\Delta_{H,2}(a, b; t)\}$ are (up to a constant) equivalent to each other in law and to fractional Brownian motion $B_H(\cdot)$; this is a direct consequence of Theorem 1.3.3. However, the situation is different when $\alpha < 2$.

Theorem 3.6.3 [CamMae89, SamTaq89] *Let $0 < H < 1$, $0 < \alpha < 2$,*

$H \neq 1/\alpha$, and let a, a', b, b' be real numbers satisfying $|a| + |b| > 0$ and $|a'| + |b'| > 0$. Then we have that

$$\left\{ \frac{1}{C_H(a,b)} \Delta_{H,\alpha}(a,b;t) \right\} \stackrel{d}{=} \left\{ \frac{1}{C_H(a',b')} \Delta_{H,\alpha}(a',b';t) \right\}, \qquad (3.6.2)$$

with

$$C_H(a,b) = \left\{ \int_{-\infty}^{\infty} \left| a\left\{ (1-v)_+^{H-1/\alpha} - (-v)_+^{H-1/\alpha} \right\} \right. \right.$$

$$\left. \left. + b\left\{ (1-v)_-^{H-1/\alpha} - (-v)_-^{H-1/\alpha} \right\} \right|^{\alpha} dv \right\}^{1/\alpha}$$

if and only if

  (i) $a = a' = 0$ or

 (ii) $b = b' = 0$ or

(iii) $aa'bb' \neq 0$ and $a/a' = b/b'$.

This theorem is proved for $\alpha \in (1,2)$ in [CamMae89] and then for any $\alpha \in (0,2)$ in [SamTaq89]. See also Theorem 7.4.5 in [SamTaq94].

Another generalization of fractional Brownian motion to the case $0 < \alpha < 2$ is *harmonizable fractional stable motion*.

Example 3.6.2 [CamMae89]   Let $0 < H < 1$, $0 < \alpha < 2$. *Harmonizable fractional stable motion* $\{\Theta_{H,\alpha}(t), t \geq 0\}$ is defined by

$$\Theta_{H,\alpha}(t) = \int_{-\infty}^{\infty} \frac{e^{it\lambda} - 1}{i\lambda} |\lambda|^{1-H-1/\alpha} d\tilde{M}_{\alpha}(\lambda),$$

where $\tilde{M}_{\alpha}$ is a complex rotationally invariant α-stable Lévy process.

Remark 3.6.1   When $\alpha = 2$, $\{\Theta_{H,2}(t)\}$ yields another representation of fractional Brownian motion.

Theorem 3.6.4 [CamMae89]   *The harmonizable fractional stable motion $\{\Theta_{H,\alpha}(t)\}$ is H-ss, si, and a rotationally invariant complex α-stable process.*

In Theorem 3.6.1 we claimed that, if $\{X(t)\}$ is H-ss, si, SαS, then $H \leq \max(1, 1/\alpha)$. A relevant question then is whether there exist H-ss, si, SαS processes for any given pair (H, α) satisfying $0 < H \leq \max(1, 1/\alpha)$, $0 < \alpha \leq 2$. For $0 < H < 1$, $0 < \alpha \leq 2$, linear fractional stable motions (fractional Brownian motion when $\alpha = 2$) are such examples. For $H = 1$, the process $\{X(t) = \xi t\}$, where ξ is a SαS random variable, is such a process. For $H > 1$ and $H = 1/\alpha$, stable Lévy processes yield examples. For

$1 < H < 1/\alpha$, (necessarily $\alpha < 1$), the following examples can be given. Examples can be constructed for any $1/2 < H < 1/\alpha$, $0 < \alpha < 2$.

Example 3.6.3 [Har82, KonMae91a] (Sub-stable processes) Let $0 < \alpha < 2$, $\alpha < \beta \le 2$, $\{Y(t)\}$ be a $S\beta S$ Lévy process and ξ be an (α/β)-stable positive random variable. Moreover suppose that $\{Y(t)\}$ and ξ are independent. Define

$$X(t) = \xi^{1/\beta} Y(t).$$

Then $\{X(t)\}$ is $1/\beta$-ss, si, $S\alpha S$, where $1/2 \le 1/\beta < 1/\alpha, 0 < \alpha < 2$.

Proof. We have, using Fubini's theorem, that

$$E\left[\exp\left\{i\theta \sum_{k=1}^{n} a_k X(t_k)\right\}\right] = E_\xi E_Y\left[\exp\left\{i\theta \xi^{1/\beta} \sum_{k=1}^{n} a_k Y(t_k)\right\}\right]$$

$$= E_\xi\left[\exp\left\{-c\left|\theta \xi^{1/\beta}\right|^\beta\right\}\right],$$

where $c = c(a_1, ..., a_n, t_1, ..., t_n)$ and E_ξ and E_Y are expectations with respect to ξ and Y, respectively. The above is equal to

$$= E_\xi\left[\exp\{-c|\theta|^\beta \xi\}\right]$$

$$= \exp\{-c'|\theta^\beta|^{\alpha/\beta}\}$$

$$= \exp\{-c'|\theta|^\alpha\},$$

where we have used $E_\xi[\exp\{-z\xi\}] = \exp\{-z^{\alpha/\beta}\}$. Thus $\{X(t)\}$ is $S\alpha S$. The fact that $\{X(t)\}$ is $1/\beta$-ss, si follows from the same property of $\{Y(t)\}$. $\quad\square$

Because of the above, we can restate Theorem 3.6.1 as follows.

Theorem 3.6.5 *Let $0 < \alpha \le 2$. A necessary and sufficient condition for the existence of H-ss, si, $S\alpha S$ processes is that $0 < H \le \max(1, 1/\alpha)$.*

We now turn to $1/\alpha$-ss, si, α-stable processes. As we have seen, the additional assumption of independence of increments yields an α-stable Lévy process. Then the problem is whether $1/\alpha$-ss, si, α-stable processes are necessarily α-stable Lévy processes, without the additional assumption of the independence of increments.

Theorem 3.6.6 *If $\alpha = 2$ or $0 < \alpha < 1$, then $1/\alpha$-ss, si, α-stable processes* *are necessarily α-stable Lévy processes. If* $1 \le \alpha < 2$, *then there exist* $1/\alpha$-*ss, si, α-stable processes other than α-stable Lévy processes.*

Proof. (i) When $\alpha = 2$, only Brownian motion has such a property. In general, an H-ss, si, 2-stable process is necessarily fractional Brownian motion $\{B_H(t)\}$ (see Theorem 1.3.3) and $\{B_H(t)\}$ with $H = 1/2$ is Brownian motion.

(ii) For $0 < \alpha < 1$, see Theorem 7.5.4 in [SamTaq94].

(iii) When $\alpha = 1$, consider $X(t) = \xi t$, where ξ is a 1-stable random variable. This is obviously a 1-ss, si, 1-stable process but not a 1-stable Lévy process, because it does not have independent increments.

(iv) For $1 < \alpha < 2$, see Examples 3.6.4 and 3.6.5 below. □

Example 3.6.4 (Sub-Gaussian processes) Let $0 < \alpha < 2$, and let ξ be an $\alpha/2$-stable positive random variable and $\{Y(t)\}$ a mean zero Gaussian process independent of ξ. Put $X(t) = \xi^{1/2} Y(t)$. This process is called a sub-Gaussian process. A calculation of its characteristic function shows that $\{X(t)\}$ is an α-stable process as in Example 3.6.3. More precisely, let $R(s, t) = E[Y(s)Y(t)]$ and use that $E[\exp\{-z\xi\}] = \exp\{-z^{\alpha/2}\}$. Then

$$E\left[\exp\left\{i\theta \sum_{k=1}^{n} a_k X(t_k)\right\}\right] = E\left[\exp\left\{i\theta \sum_{k=1}^{n} a_k \xi^{1/2} Y(t_k)\right\}\right]$$

$$= E_\xi E_Y\left[\exp\left\{i\theta \xi^{1/2} \sum_{k=1}^{n} a_k Y(t_k)\right\}\right]$$

$$= E_\xi\left[\exp\left\{-\frac{1}{2}|\theta|^2 \xi \sum_{k,j=1}^{n} a_k a_j R(t_k, t_j)\right\}\right]$$

$$= \exp\left\{-|\theta|^\alpha \left[\frac{1}{2} \sum_{k,j=1}^{n} a_k a_j R(t_k, t_j)\right]^{\alpha/2}\right\}. \quad (3.6.3)$$

Suppose that $1 < \alpha < 2$. If we take a fractional Brownian motion $B_{1/\alpha}$ as the Gaussian process $\{Y(t)\}$, then we see that $\{X(t)\}$ is $1/\alpha$-ss, si. However, as will be shown in the following lemma, $\{X(t)\}$ cannot have independent increments, and hence, it is not an α-stable Lévy process.

Lemma 3.6.1 *Nondegenerate, jointly sub-Gaussian random variables cannot be independent.*

Proof. Let X_1 and X_2 be nondegenerate jointly sub-Gaussian such that $X_1 =$

$\xi^{1/2}Y_1$, $X_2 = \xi^{1/2}Y_2$, Y_1 and Y_2 are jointly Gaussian, and $r_{ij} = E[Y_iY_j]$. Then by (3.6.3)

$$E[\exp\{i(a_1X_1 + a_2X_2)\}] = \exp\left\{-\left[\frac{1}{2}\left(a_1^2r_{11} + 2a_1a_2r_{12} + a_2^2r_{22}\right)\right]^{\alpha/2}\right\}.$$

On the other hand, if X_1 and X_2 are independent,

$$E[\exp\{i(a_1X_1 + a_2X_2)\}] = E[\exp\{ia_1X_1\}]E[\exp\{ia_2X_2\}]$$

$$= \exp\left\{-\left(\frac{1}{2}a_1^2r_{11}\right)^{\alpha/2} - \left(\frac{1}{2}a_2^2r_{22}\right)^{\alpha/2}\right\}.$$

Thus, we have for all a_1, $a_2 \in \mathbb{R}$,

$$\left(a_1^2r_{11} + 2a_1a_2r_{12} + a_2^2r_{22}\right)^{\alpha/2} = \left(a_1^2r_{11}\right)^{\alpha/2} + \left(a_2^2r_{22}\right)^{\alpha/2},$$

meaning that

$$x^2 + 2\rho x + 1 = (|x|^\alpha + 1)^{2/\alpha} \qquad \text{for all } x \in \mathbb{R},$$

where $\rho = r_{12}(r_{11}r_{22})^{-1/2} \in [-1, 1]$. However, this is impossible. □

Example 3.6.5 [KasMaeVer88] (Log-fractional stable motion)   Define

$$\Delta_{\log}(t) = \int_{-\infty}^{\infty} \log\left|\frac{t-u}{u}\right| dZ_\alpha(u).$$

This process is well-defined for $1 < \alpha \le 2$, and $S\alpha S$ by Theorem 3.5.2. It is also easily checked to be $1/\alpha$-ss, si. When $\alpha = 2$, it is a Brownian motion. Indeed, it was shown in Theorem 1.3.3 that a $1/2$-ss, si Gaussian process is a Brownian motion. This representation for Brownian motion yields another example for Remark 3.5.2. However, when $\alpha < 2$, it cannot have independent increments. Indeed, consider disjoint increments of $\{\Delta_{\log}(t)\}$. The integrands of these stable integrals do not have disjoint supports. By Theorem 3.5.3, these increments are not independent, and hence the process is not an α-stable Lévy process.

For a further interesting example of a process which is H-ss, si with $H > 1/2$, see Theorem 3.8.2.

3.7 LONG-RANGE DEPENDENCE (II)

The dependence between increments of selfsimilar processes with stationary increments, not having finite variances, cannot be measured by correlations. Taqqu and his collaborators used a notion measuring the dependence of increments of selfsimilar *stable* processes with stationary increments. (See [SamTaq94, Sections 2.10 and 4.7].)

Definition 3.7.1 *Let $0 < \alpha \leq 2$ and let (X, Y) be a two-dimensional $S\alpha S$ random vector, and suppose that $X \sim S\alpha S(\sigma_X^\alpha)$, $Y \sim S\alpha S(\sigma_Y^\alpha)$ and $X - Y \sim S\alpha S(\sigma_{X-Y}^\alpha)$. The "codifference" $\tau_{X,Y}$ is defined by*

$$\tau_{X,Y} = \sigma_X^\alpha + \sigma_Y^\alpha - \sigma_{X-Y}^\alpha.$$

Remark 3.7.1 If X and Y are independent, then $\tau_{X,Y} = 0$; indeed

$$\sigma_{X-Y}^\alpha |\theta|^\alpha = -\log E\left[e^{i\theta(X-Y)}\right]$$

$$= -\log E\left[e^{i\theta X}\right] - \log E\left[e^{-i\theta Y}\right] = (\sigma_X^\alpha + \sigma_Y^\alpha)|\theta|^\alpha.$$

Remark 3.7.2 When $\alpha = 2$, $\tau_{X,Y} = \mathrm{Cov}(X, Y)$. Indeed, when $\alpha = 2$, $\sigma_X^2 = \frac{1}{2}\mathrm{Var}X$. Thus,

$$\tau_{X,Y} = \frac{1}{2}(\mathrm{Var}X + \mathrm{Var}Y - \mathrm{Var}(X - Y))$$

$$= \frac{1}{2}\left(E[X^2] + E[Y^2] - E[(X - Y)^2]\right)$$

$$= \frac{1}{2}2E[XY] = \mathrm{Cov}(X, Y).$$

Here we have used, without loss of generality, that the means are zero.

The above definition can be used only for symmetric stable random variables, but the following characteristic function based notion can be used for general random variables. Let $\{Y(t)\}$ be a stationary process. Define

$$U(\theta_1, \theta_2; t) = E\left[e^{i(\theta_1 Y(t) + \theta_2 Y(0))}\right] - E\left[e^{i\theta_1 Y(t)}\right]E\left[e^{i\theta_2 Y(0)}\right]$$

and

$$I(\theta_1, \theta_2; t) = -\log E\left[e^{i(\theta_1 Y(t) + \theta_2 Y(0))}\right] + \log E\left[e^{i\theta_1 Y(t)}\right] + \log E\left[e^{i\theta_2 Y(0)}\right].$$

We have that

$$U(\theta_1, \theta_2; t) = K(\theta_1, \theta_2)\left(e^{-I(\theta_1, \theta_2; t)} - 1\right),$$

where

$$K(\theta_1, \theta_2) = E\left[e^{i\theta_1 Y(t)}\right]E\left[e^{i\theta_2 Y(0)}\right] = E\left[e^{i\theta_1 Y(0)}\right]E\left[e^{i\theta_2 Y(0)}\right].$$

If $I(\theta_1, \theta_2; t) \to 0$ as $t \to \infty$, then

$$U(\theta_1, \theta_2; t) \sim -K(\theta_1, \theta_2)I(\theta_1, \theta_2; t),$$

and thus checking dependence by U is the same as by I. For simplicity, we write $\tau(t) = \tau_{Y(t), Y(0)}$. If $Y(t)$ and $Y(0)$ are jointly $S\alpha S$ distributed, then $\tau(t) = -I(1, -1; t)$.

Remark 3.7.3 If $I(\theta_1, \theta_2; t) \to 0$ as $t \to \infty$, then the stationary process $\{Y(t)\}$ is mixing, namely

$$\lim_{t \to \infty} |P(A \cap B) - P(A)P(B)| = 0$$

for any $A \in \sigma\{Y(s), \ s \le 0\}$, $B \in \sigma\{Y(s), \ s \ge t\}$ (see [Mar70]).

For an application of the above, let us calculate the dependence structure of a linear fractional stable motion, discussed in Example 3.6.1. For simplicity take

$$X(t) = \int_{\mathbb{R}} \left(|t - u|^{H - 1/\alpha} - |u|^{H - 1/\alpha}\right) dZ_\alpha(u),$$

and consider the stationary increments

$$Y_n = X(n + 1) - X(n).$$

Theorem 3.7.1 [AstLevTaq91]
  (i) *If* $0 < \alpha \le 1$, $0 < H < 1$ *or* $1 < \alpha < 2$, $1 - 1/\alpha(\alpha - 1) < H < 1$, $H \ne 1/\alpha$, *then*

$$I(\theta_1, \theta_2; n) \sim B(\theta_1, \theta_2)n^{\alpha H - \alpha},$$

$$\tau(n) \sim -B(1, -1)\, n^{\alpha H - \alpha}. \tag{3.7.1}$$

  (ii) *If* $1 < \alpha \le 2$, $0 < H < 1 - 1/\alpha(\alpha - 1)$, *then*

$$I(\theta_1, \theta_2; n) \sim F(\theta_1, \theta_2)n^{H - 1/\alpha - 1},$$

$$\tau(n) \sim -F(1, -1)n^{H - 1/\alpha - 1}.$$

Here $B(\theta_1, \theta_2)$ *and* $F(\theta_1, \theta_2)$ *are positive constants which can be calculated explicitly.*

Remark 3.7.4 When $\alpha = 2$, it follows from (3.2.1) with $H \ne 1/2$ that the covariance $r(n)$ satisfies

$$r(n) \sim Cn^{2H - 2},$$

for some $C \neq 0$. This corresponds to (3.7.1) with $\alpha = 2$. The statement (ii) above shows, however, that this analogy does not hold when H is small.

When one wants to show that two $S\alpha S$ stationary processes are different, we may be able to use $\tau(t)$ or I. If their asymptotics are different, they are different processes. There are some $S\alpha S$ stationary processes where the $\tau(t)$ are the same, but the $I(\theta_1, \theta_2; t)$ are different for some θ_1, θ_2. In this sense, even in the case of $S\alpha S$ processes, I is more useful than τ.

Another interesting idea for measuring the dependence was given by Heyde and Yang [HeyYan97]. Suppose $\{X_j\}$ is a stationary sequence, centered to have zero mean when the mean is finite. They say that the sequence is long-range dependent in the sense of sample Allen variance (LRD(SAV), for short) if

$$\frac{\left(\sum_{j=1}^{m} X_j\right)^2}{\sum_{j=1}^{m} X_j^2} \longrightarrow \infty$$

in probability as $m \to \infty$. This definition does not require any moments to exist. The following result was obtained.

Theorem 3.7.2 [HeyYan97] *Let $\{X(t)\}$ be H-ss, si, and suppose that $E[|X(1)|^p] < \infty$ for some $0 < p < 2$. Then LRD(SAV) holds if $H > 1/p$.*

This corresponds to the case $H > 1/2$ in Section 3.2.

3.8 LIMIT THEOREMS (III)

We now concentrate on two selfsimilar processes discussed in Section 3.6 and give some limit theorems. The two processes are:

$$X_1(t) = \int_{-\infty}^{\infty} \left(|t - u|^{H-1/\alpha} - |u|^{H-1/\alpha}\right) dZ_\alpha(u), \ t \geq 0, \ 0 < H < 1, \ H \neq \frac{1}{\alpha},$$

and

$$X_2(t) = \int_{-\infty}^{\infty} \log\left|\frac{t-u}{u}\right| dZ_\alpha(u), \ t \geq 0, \ 1 < \alpha \leq 2.$$

Suppose $\{X_j, j \in \mathbb{Z}\}$ are independent and identically distributed symmetric random variables satisfying

$$\frac{1}{n^{1/\alpha}} \sum_{j=1}^{n} X_j \xrightarrow{d} Z_\alpha(1). \tag{3.8.1}$$

Take δ such that $1/\alpha - 1 < \delta < 1/\alpha$, and define a stationary sequence

$$Y_k = \sum_{j \in \mathbb{Z}} c_j X_{k-j}, \qquad k = 1, 2, \ldots,$$

where

$$c_j = \begin{cases} 0, & \text{if } j = 0, \\ j^{-\delta-1}, & \text{if } j > 0, \\ -|j|^{-\delta-1}, & \text{if } j < 0. \end{cases}$$

We can easily see that the infinite series $\{Y_k\}$ is well defined for each k and Y_k does not have finite variance unless $\alpha = 2$. Define further for $H = 1/\alpha - \delta$,

$$W_n(t) := \frac{1}{n^H} \sum_{k=1}^{[nt]} Y_k. \tag{3.8.2}$$

Theorem 3.8.1 *Under the above assumptions,*

$$W_n(t) \overset{d}{\Rightarrow} \begin{cases} \dfrac{1}{|\delta|} X_1(t) & \text{when } \delta \neq 0, \\ X_2(t) & \text{when } \delta = 0. \end{cases}$$

Remark 3.8.1 If $\delta < 0$ (necessarily $\alpha > 1$), then $H = 1/\alpha - \delta > 1/\alpha$. Thus the normalization n^H in (3.8.2) grows much faster than $n^{1/\alpha}$ in (3.8.1), the case of partial sums of independent random variables. This explains why $\{Y_k\}$ exhibits long-range dependence.

We give an outline of the proof of Theorem 3.8.1.

Step 1. For $m \in \mathbb{Z}$ and $t \geq 0$, define

$$c_m(t) = \sum_{j=1-m}^{[t]-m} c_j,$$

where $\sum_{j=1-m}^{-m}$ means 0. Then we have that

$$W_n(t) = n^{-H} \sum_{m \in \mathbb{Z}} c_m(nt) X_m.$$

Step 2. For any $t_1, \ldots, t_p \geq 0$ and $\theta_1, \ldots, \theta_p \in \mathbb{R}$,

$$\sum_{m \in \mathbb{Z}} \left| n^{-H} \sum_{j=1}^{p} \theta_j c_m(nt_j) \right|^{\alpha}$$

$$\rightarrow \begin{cases} \int_{-\infty}^{\infty} \left| \dfrac{1}{|\delta|} \sum_{j=1}^{p} \theta_j \left(|t_j - u|^{-\delta} - |u|^{-\delta} \right) \right|^{\alpha} du & \text{when } \delta \neq 0, \\[2em] \int_{-\infty}^{\infty} \left| \sum_{j=1}^{p} \theta_j \log \dfrac{|t_j - u|}{|u|} \right|^{\alpha} du & \text{when } \delta = 0. \end{cases}$$

Step 3. Denote the characteristic function of X_1 by $\lambda(\theta)$, $\theta \in \mathbb{R}$. Then we have that

$$\log \lambda(\theta) \sim -|\theta|^{\alpha} \qquad \text{as } \theta \to 0,$$

[MaeMas94].
Also

$$\lim_{n \to \infty} n^{-H} \sup_{m} c_m(n) = 0,$$

[Mae83].
Step 4. We have that

$$I_n := E\left[\exp\left\{ n^{-H} \sum_{j=1}^{p} \theta_j W_n(t) \right\} \right]$$

$$= E\left[\exp\left\{ n^{-H} \sum_{j=1}^{p} \theta_j \sum_{m \in \mathbb{Z}} c_m(nt_j) X_m \right\} \right]$$

$$= E\left[\prod_{m \in \mathbb{Z}} \lambda\left(n^{-H} \sum_{j=1}^{p} \theta_j c_m(nt_j) \right) \right]$$

and, by Steps 2 and 3,

$$\lim_{n \to \infty} I_n = \lim_{n \to \infty} E\left[\prod_{m \in \mathbb{Z}} \lambda\left(n^{-H} \sum_{j=1}^{p} \theta_j c_m(nt_j) \right) \right]$$

$$= \lim_{n \to \infty} E\left[\exp\left\{ \sum_{m \in \mathbb{Z}} \log \lambda\left(n^{-H} \sum_{j=1}^{p} \theta_j c_m(nt_j) \right) \right\} \right]$$

$$= \begin{cases} E\left[\exp\left\{-\int_{-\infty}^{\infty}\left|\frac{1}{|\delta|}\sum_{j=1}^{p}\theta_j\left(|t_j-u|^{-\delta}-|u|^{-\delta}\right)\right|^{\alpha}du\right\}\right] & \text{when } \delta \neq 0 \\ \\ E\left[\exp\left\{-\int_{-\infty}^{\infty}\left|\sum_{j=1}^{p}\theta_j\log\left(\frac{|t_j-u|}{|u|}\right)\right|^{\alpha}du\right\}\right] & \text{when } \delta = 0 \end{cases}$$

$$= \begin{cases} E\left[\exp\left\{i\frac{1}{|\delta|}\sum_{j=1}^{p}\theta_j X_1(t_j)\right\}\right] & \text{when } \delta \neq 0 \\ \\ E\left[\exp\left\{i\sum_{j=1}^{p}\theta_j X_2(t_j)\right\}\right] & \text{when } \delta = 0, \end{cases}$$

where at the last stage, we have used that, for $f \in L^{\alpha}(\mathbb{R})$ and $X_{\alpha} = \int_{-\infty}^{\infty} f(u)\, dZ_{\alpha}(u)$:

$$E\left[e^{i\theta X_{\alpha}}\right] = \exp\left\{-|\theta|^{\alpha}\int_{-\infty}^{\infty}|f(u)|^{\alpha}\, du\right\}.$$

(See Theorem 3.5.1.) The above Step 4 gives us the conclusion. □

Kesten and Spitzer [KesSpi79] constructed an interesting class of ss, si processes as a limit of *random walks in random scenery*, where the limiting process is expressed as a stable-integral process with a random integrand. Let $\{Z_{\alpha}(t), t \in \mathbb{R}\}$ be a symmetric α-stable Lévy process $(0 < \alpha \leq 2)$ and $\{Z_{\beta}(t), t \in \mathbb{R}\}$ a symmetric β-stable Lévy process $(1 < \beta \leq 2)$ independent of $\{Z_{\alpha}(t)\}$. Let $L_t(x)$ be the local time of $\{Z_{\beta}(t)\}$, that is

$$L_t(x) = \lim_{\varepsilon \downarrow 0} \frac{1}{4\varepsilon}\int_0^t I\left[Z_{\beta}(s) \in (x-\varepsilon, x+\varepsilon)\right] ds,$$

which is known to exist as an almost sure limit if $1 < \beta \leq 2$ [Boy64]. Then we can define

$$X(t) = \int_{-\infty}^{\infty} L_t(x)\, dZ_{\alpha}(x)$$

and $\{X(t),\ t \geq 0\}$ is H-ss, si, with $H = 1 - 1/\beta + 1/\alpha\beta(> 1/2)$, since $\{Z_{\alpha}(t)\}$ is a semimartingale.

A limit theorem for this process $\{X(t)\}$ is given as follows. Let $\{S_n, n \geq 0\}$ be an integer-valued random walk with mean 0 and $\{\xi(j), j \in \mathbb{Z}\}$ be a sequence of symmetric independent and identically distributed random variables, independent of $\{S_n\}$ such that

$$\frac{1}{n^{1/\alpha}} \sum_{j=1}^{n} \xi(j) \overset{d}{\to} Z_\alpha(1) \qquad \text{and} \qquad \frac{1}{n^{1/\beta}} S_n \overset{d}{\to} Z_\beta(1).$$

The new stationary sequence $\{\xi(S_k)\}$, which is a random walk in random scenery, is strongly dependent.

Theorem 3.8.2 [KesSpi79]   *Under the above assumptions, we have that*

$$\frac{1}{n^H} \sum_{k=1}^{[nt]} \xi(S_k) \overset{d}{\Rightarrow} \int_{-\infty}^{\infty} L_t(x) dZ_\alpha(x).$$

Chapter Four

Fractional Brownian Motion

Although we have mentioned fractional Brownian motion in Section 1.3, we discuss this important process in more detail in this chapter.

4.1 SAMPLE PATH PROPERTIES

When two stochastic processes $\{X(t)\}$ and $\{Y(t)\}$ satisfy $P\{X(t) = Y(t)\} = 1$ for all $t \geq 0$, we say that one is a modification of the other. It is well known that Brownian motion has a modification, the sample paths of which are continuous almost surely, but sample paths of any modification are nowhere differentiable. As it turns out, these facts remain true for fractional Brownian motion.

We say that a stochastic process $\{X(t), 0 \leq t \leq T\}$ is Hölder continuous of order $\gamma \in (0, 1)$ if

$$P\left\{\omega \in \Omega, \sup_{\substack{0 < t - s < h(\omega) \\ s, t \in [0, T]}} \frac{|X(t, \omega) - X(s, \omega)|}{|t - s|^{\gamma}} \leq \delta\right\} = 1,$$

where h is an almost surely positive random variable and $\delta > 0$ is an appropriate constant.

Lemma 4.1.1 (A general version of Kolmogorov's criterion) *If a stochastic process $\{X(t)\}$ satisfies*

$$E\left[|X(t) - X(s)|^{\delta}\right] \leq C|t - s|^{1+\varepsilon}, \qquad \forall t, s, \qquad (4.1.1)$$

for some $\delta > 0$, $\varepsilon > 0$ and $C > 0$, then $\{X(t)\}$ has a modification, the sample paths of which are Hölder continuous of order $\gamma \in [0, \varepsilon/\delta)$.

For a proof, see for instance [KarShr91, p. 53].

Theorem 4.1.1 *Fractional Brownian motion $\{B_H(t)\}$, $0 < H < 1$, has a modification, the sample paths of which are Hölder continuous of order $\beta \in [0, H)$.*

Proof. Choose $0 < \gamma < H$. Then we have by selfsimilarity and stationary increments of $\{B_H(t)\}$,

$$E\left[|B_H(t) - B_H(s)|^{1/\gamma}\right] = E\left[|B_H(|t - s|)|^{1/\gamma}\right]$$

$$= |t - s|^{H/\gamma} E\left[|B_H(1)|^{1/\gamma}\right].$$

Then (4.1.1) is satisfied with $\delta = 1/\gamma$ and $\varepsilon = H/\gamma - 1$. Thus, there exists a modification which is Hölder continuous of order $\beta < (H/\gamma - 1)\gamma = H - \gamma$. Since γ can be arbitrarily small, the result follows. $\square$

Theorem 4.1.2 [Ver85] *Suppose* $\{X(t)\}$ *is H-ss, si. If* $H \leq 1$ *and* $P\{X(t) = tX(1)\} = 0$, *then the sample paths of* $\{X(t)\}$ *have infinite variation, almost surely, on all compact intervals.*

Remark 4.1.1 Theorem 4.1.2 does not hold for $H > 1$, as can be seen from the α-stable Lévy process with $\alpha < 1$.

Corollary 4.1.1 *Sample paths of fractional Brownian motion have nowhere bounded variation.*

Since fractional Brownian motion is H-ss, si, $0 < H < 1$, Corollary 4.1.1 is a direct consequence of Theorem 4.1.2.

Define a partition of $[0, T]$ as the set of pairs of consecutive dividing points, namely

$$\Delta_n = \{(t_0, t_1), (t_1, t_2), \ldots, (t_{k-1}, t_k) : 0 = t_0 < t_1 < \cdots < t_n = T\}.$$

Also define $|\Delta_n| = \max\{|t_j - t_{j-1}|, 1 \leq j \leq n\}$ and consider sequences of partitions Δ_n with $\lim_{n\to\infty} |\Delta_n| = 0$. The following result is due to Rogers [Rog97].

Lemma 4.1.2 *Fix* $p > 0$. *Then, using the above notation,*

$$V_{n,p} := \sum_{t_j \in \Delta_n} \left|B_H(t_{j+1}) - B_H(t_j)\right|^p$$

$$\to \begin{cases} 0 & \text{if } pH > 1 \\ +\infty & \text{if } pH < 1 \end{cases}$$

in the sense of convergence in probability, as $n \to \infty$.

Hence if $p < 1/H$, then $V_p := \lim_{n\to\infty} V_{n,p}$ is almost surely infinite, possibly along a subsequence if necessary.

Theorem 4.1.3 *Sample paths of fractional Brownian motion $\{B_H(t)\}$ are almost surely nowhere locally Hölder continuous of order γ for $\gamma > H$ in the sense that there is no interval on which they are Hölder continuous of order γ.*

Proof. Suppose that for some $\gamma > H$,

$$|B_H(t) - B_H(s)| \leq C|t - s|^\gamma$$

for all $t, s \in [0, T]$ and choose p such that $H < 1/p < \gamma$. Then

$$V_{n,p} = \sum_{(u,v)\in\Delta_n} |B_H(u) - B_H(v)|^p \leq C^p T |\Delta_n|^{p\gamma-1},$$

where the left hand side diverges by Lemma 4.1.2 and the right hand side converges to zero. This completes the proof. $\square$

Nowhere differentiability of sample paths of fractional Brownian motion is shown as a corollary of a theorem in [KawKon71], where the authors proved nowhere differentiability of sample paths for a class of Gaussian processes including fractional Brownian motion.

4.2 FRACTIONAL BROWNIAN MOTION FOR $H \neq 1/2$ IS NOT A SEMIMARTINGALE

In this section, we show that fractional Brownian motion, for $H \neq 1/2$, is not a semimartingale. Though this result has been known for a long time in the more mathematical literature, in more applied publications, especially in finance, the consequences of this result resurfaced only fairly recently. A proof for $H > 1/2$ can be found in [Lin95]. The case $0 < H < 1$ ($H \neq 1/2$) can be found in [Rog97]. For an early discussion and proof, see [LipShi89, p. 300, Example 2]. Of course, Brownian motion ($\{B_H(t)\}$ with $H = 1/2$) is a martingale. This allows us to construct the so-called Itô calculus with respect to Brownian motion. On the other hand, the role of fractional Brownian motion has been increasing. As a consequence, stochastic integrals with respect to fractional Brownian motion are needed. The non-semimartingale property implies that the "classical" construction and properties do not hold. The classical theory therefore has to be adapted. We shall return to this point in Section 4.3.

Theorem 4.2.1 $\{B_H(t)\}$, $H \neq 1/2$, is not a semimartingale.

Proof. (i) Case $1/2 < H < 1$ [LipShi89, pp. 299–300]. We first show that if $1/2 < H < 1$, then the quadratic variation of $\{B_H(t)\}$ is a zero process.

Consider a sequence of partitions of $[0, T]$, Δ_n with $\lim_{n \to \infty} |\Delta_n| = 0$. By selfsimilarity, we can assume that $T = 1$ by scaling. We have that, if $E[B_H(1)^2] = 1$, then

$$E\left[\sum_{(u,v) \in \Delta_n} |B_H(u) - B_H(v)|^2\right] = \sum_{(u,v) \in \Delta_n} (v - u)^{2H}$$

$$= |\Delta_n|^{2H-1} \sum_{(u,v) \in \Delta_n} (v - u) \left(\frac{v - u}{|\Delta_n|}\right)^{2H-1}$$

$$\leq |\Delta_n|^{2H-1} \to 0,$$

as $n \to \infty$, since $2H > 1$. If $\{B_H(t)\}$ were to be a semimartingale, it would have a Doob–Meyer decomposition $B_H(t) = M(t) + V(t)$, where $\{M(t)\}$ is a continuous local martingale and $\{V(t)\}$ is a finite variation process with $M(0) = V(0) = 0$. Denote by $[Z, Z]_t$ the quadratic variation process of a semimartingale $\{Z(t)\}$. Then by the above observation,

$$0 = [B_H, B_H]_t = [M, M]_t.$$

By the Burkholder–Gundy–Davis inequality, $\{M(t)\}$ is itself a zero process, and hence $\{B_H(t)\} = \{V(t)\}$ has finite variation, which contradicts Corollary 4.1.1.

(ii) Case $0 < H < 1/2$ [Rog97]. We have that

$$I_n := \sum_{j=1}^n \left|B_H\left(\frac{j}{n}\right) - B_H\left(\frac{j-1}{n}\right)\right|^2 \overset{d}{=} \frac{1}{n^{2H}} \sum_{j=1}^n |B_H(j) - B_H(j-1)|^2$$

$$= n^{1-2H} \frac{1}{n} \sum_{j=1}^n |B_H(j) - B_H(j-1)|^2.$$

Hence by the ergodic theorem,

$$\frac{1}{n} \sum_{j=1}^n |B_H(j) - B_H(j-1)|^2 \to E\left[|B_H(1)|^2\right] > 0$$

almost surely, and hence since $0 < H < 1/2$,

$$P\{I_n \leq x\} \to 0, \qquad \forall x \in \mathbb{R}, \tag{4.2.1}$$

so that $I_n \to \infty$ in law and in probability as $n \to \infty$. If $\{B_H(t)\}$ were a semimartingale, its quadratic variation must be finite almost surely, but this contradicts (4.2.1). $\square$

4.3 STOCHASTIC INTEGRALS WITH RESPECT TO FRACTIONAL BROWNIAN MOTION

Due to the popularity of selfsimilar processes in various applications, the demand for a stochastic calculus based upon such processes has increased considerably over the last couple of years. Especially SDEs driven by fractional Brownian motion are high in demand, in particular in physics, telecommunication and finance. As already stated above, the non-semimartingale property of B_H for $H \neq 1/2$ (see Theorem 4.2.1) makes the standard construction fail. Specific choices with respect to the definition of

$$\int_0^t \psi(s) \, dB_H(s)$$

have to be made. More precisely, we cannot obtain a fully satisfactory theory when integrating over all predictable processes (predictable with respect to the filtration generated by fractional Brownian motion). This is a consequence of the famous Bichteler–Dellacherie theorem; see [DelMey80, VIII.4] and [Bic81]. Because of the sample path properties of fractional Brownian motion (see Theorem 4.1.1), for $1/2 < H < 1$, fractional Brownian motion has smoother sample paths than Brownian motion and hence it will be easier to construct a stochastic integral. Indeed, in this case we can follow a pathwise Riemann–Stieltjes construction. For $0 < H < 1/2$, sample paths of fractional Brownian motion are more irregular than those of Brownian motion and therefore other constructions have to be followed. For simulated sample paths in these cases, see Section 7.4. The most important constructions (for general H) to be found in the literature either restrict ψ to specific classes of functions, use pathwise integration or base a definition on Malliavin calculus. In each of these approaches, a version of the Itô formula is deduced which allows for a calculus of SDEs. As far as we are aware, no general consensus on the "right" approach exists. In view of the sample path (ir)regularity discussed above, for $1/2 < H < 1$, the much easier pathwise approach has to be favored. Nevertheless, the recent literature abounds with different constructions, even in this case. For applications to mathematical finance, this is particularly frustrating because the notion of stochastic integral is intimately linked to concepts like arbitrage, completeness, strategies, etc. Starting from a Black–Scholes type of world for the price $S(t)$ of a financial instrument at time t,

$$dS(t) = S(t)\big(\mu \, dt + \sigma \, dB_{1/2}(t)\big),$$

a naive "replacement" of Brownian motion $B_{1/2}$ by a general fractional Brownian motion B_H, $H \neq 1/2$, leading to a so-called fractional Black–Scholes model

$$dS(t) = S(t)\big(\mu \, dt + \sigma \, dB_H(t)\big), \tag{4.3.1}$$

has to be treated with care.

In [Shi98], Shiryaev takes $f \in C^2$ and uses a second order Taylor expansion on f in order to rewrite $f(B_H(t)) - f(B_H(0))$ in the case where $1/2 < H < 1$. A limit argument yields an interpretation of

$$\int_0^t f'(B_H(s))\, dB_H(s),$$

leading to an Itô formula of the type

$$f(B_H(t)) - f(B_H(0)) = \int_0^t f'(B_H(s))\, dB_H(s),$$

almost surely.

Note that this formula leads to $E[\int_0^t B_H(s)\, dB_H(s)] \neq 0$ rendering the terminology "Itô type formula" questionable. Shiryaev [Shi98] further discusses the construction of a fractional Black–Scholes market.

Lin [Lin95] gives a similar construction to Shiryaev's above. He also extends the definition of the stochastic integral to càdlàg functions using a Riemann sum based approach. Namely, for

$$\psi(s) = \sum_{j=1}^n a_j 1_{(t_{j-1}, t_j]}(s),$$

where $\{t_0, \ldots, t_n\}$ defines a partition of $[0, t]$,

$$\int_0^t \psi(s)\, dB_H(s) = \sum_{j=1}^n a_j \Big(B_H(t_j) - B_H(t_j - 1)\Big).$$

An L^2 limit result establishes the existence of a stochastic integral. A similar approach is to be found in [DaiHey96]; they also discuss the fractional Black–Scholes model in their set-up. In [GriNor96] and [DunHuPas00] the above restrictions on ψ are slightly different. It is worthwhile remarking that the change of variable formulae obtained through these constructions are of the Stratonovich type, hence like the usual result for deterministic functions of bounded variation. Although this is useful, one typically ends up with properties like

$$E\left[\int_0^t \psi(s)\, dB_H(s)\right] \neq 0.$$

An excellent survey of the various results in the above approaches, comparing Stratonovich with Itô, checking whether $E[\int_0^t \psi(s)\, dB_H(s)] = 0$, and comparing and contrasting the resulting change of variable formulae is to be found in [DunHuPas00]. Using the notion of Wick products, a new type of stochastic integral of the Stratonovich type is defined, relating it with the definitions in [Lin95] and [DaiHey96].

A rather different approach (again $1/2 < H < 1$) is taken by Mikosch and Norvaiša [MikNor00]. The notion of p-variation ($0 < p < \infty$) of a real function defined on $[a, b]$, say, plays a central role here:

$$v_p(\psi) = v_p(\psi : [a, b]) = \sup_\Delta \sum_{j=1}^k \left| \psi(x_j) - \psi(x_{j-1}) \right|^p,$$

where $\Delta = \{x_0, ..., x_k\}$ defines any partition on $[a, b]$, $x_0 = a, x_k = b$. For $v_p(\psi) < \infty$, ψ is said to have bounded p-variation. The case $p = 1$ corresponds to the usual definition of bounded variation of ψ. Recall the difference between 2-variation and quadratic variation of a stochastic process. The latter is defined as the limit of the quantities $\sum_{(u,v) \in \Delta_n} |\psi(u) - \psi(v)|^2$ provided that this limit exists. Also recall that sample paths of Brownian motion have unbounded 2-variation and bounded p-variation for every $p > 2$. In the case of fractional Brownian motion $(0 < H < 1)$, B_H has bounded p-variation almost surely for any $p > 1/H$ [KawKon73]. Based on these results, the authors show that Riemann–Stieltjes integral equations driven by sample paths of fractional Brownian motion are appropriate. The allowable integrands have to have finite q-variation where $1/p + 1/q > 1$. It is easily checked that for $1/2 < H < 1$, $\int_0^t B_H(s) \, dB_H(s)$ exists pathwise; this in contrast to the case $H = 1/2$ (Brownian motion) for which, for instance, an Itô integral has to be defined. Other (integration) processes which fit the approach in [MikNor00] are general Lévy processes. The change of variable rule obtained is similar to the Itô's classical rule for Brownian motion. Applications to the Langevin equation under additive and multiplicative fractional Brownian motion noise are discussed. They also show that for the existence of a unique solution of (4.3.1) one needs that B_H has bounded p-variation for some $p < 2$. So, if $H > 1/2$, then $v_p(B_H) < \infty$ for $1/H < p < 2$, and hence the fractional Black–Scholes stochastic differential equation (4.3.1) has a unique solution. The paper contains various references for further reading.

Using Malliavin calculus, [DunHuPas00] also contains a definition of an Itô type integral $\int_0^t F(s) \, dB_H(s)$ where F is a stochastic process. In their notation, a change of variable formula of the following type is obtained:

$$f(B_H(t)) - f(B_H(0)) = (\text{Itô}) \int_0^t f'(B_H(s)) \, dB_H(s)$$

$$+ H \int_0^t s^{2H-1} f''(B_H(s)) \, ds,$$

almost surely.

It is interesting to note that this formula yields the usual Itô formula for Brownian motion when $H = 1/2$ is formally substituted.

An interesting paper discussing stochastic integration with respect to fractional Brownian motion B_H for general $0 < H < 1$ is [CarCou00]. The basic idea underlying their construction is that of regularization. Namely, an integral with respect to fractional Brownian motion is constructed through a sequence of approximating integrals. The latter are defined with respect to semimartingales. For $H > 1/4$, a natural Itô formula, previously postulated by Privault [Pri98], is obtained. The authors also compare and contrast their approach with some of those discussed above. The idea of regularization, together with applications for option pricing in fractional markets, is also discussed in Cheridito [Che00b]. Those interested in some further properties of fractional Brownian motion, including a detailed discussion on arbitrage in such markets should further consult [Che00a, Che01]. Finally, using the ideas of Carmona and Coutin [CarCou00], Alós, Mazet and Nualart [AloMazNua00] develop a stochastic calculus with respect to fractional Brownian motion with $0 < H < 1/2$. The basic tool used is Malliavin calculus. See also [PipTaq00, PipTaq01] for an interesting summary in the case $H \in (0, 1)$.

At this point, a natural question is for which general selfsimilar processes X can one define a sufficiently rich notion of stochastic integral

$$\int_0^t \psi(s) \, dX(s).$$

As we have already seen, "good" classes of integrators are fractional Brownian motion and α-stable Lévy processes. For $H > 1/2$, the pathwise approach of [MikNor00] works when X has finite p-variation with $p \in (1/H, 2)$. The problem, however, is how to check this latter property. For Gaussian processes, [KawKon73] gave a solution. For non-Gaussian processes, the calculation of p-variation properties typically becomes very difficult. However, if the process is symmetric stable, we can use the idea of representing symmetric stable processes by infinite series, which are conditionally Gaussian processes; see for instance [Ros90].

Since boundedness of p-variation follows from Hölder continuity, it is enough to check Hölder continuity of sample paths of stochastic processes. Hölder continuity for some selfsimilar processes is studied in [KonMae91b] as an application of the conditionally Gaussian representation of stable processes. From the results there, we see that the linear fractional stable motion $\Delta_{H,\alpha}$ in Example 3.6.1 has bounded p-variation for $p > 1/(H - (1/\alpha))$, where $1 < \alpha < 2$ and $1/\alpha < H < 1$. In this case we cannot find $p < 2$ for which $v_p(\Delta_{H,\alpha}) < \infty$. So, we cannot obtain a Black–Scholes model with driving process $\Delta_{H,\alpha}$.

However, the harmonizable fractional stable motion Θ_H, α in Example 3.6.2 is Hölder continuous of order $q < H$ and has bounded p-variation for $p > 1/H$. Thus, if $H > 1/2$, a Black–Scholes model with the real part of $\Theta_{H,\alpha}$ as a driving process is well defined and has a unique solution.

Finally, the above mentioned approach using regularization is useful for finding interesting practical models which at the same time have nice stochastic properties. Recall for instance the definition of fractional Brownian motion as given in Theorem 1.3.3. An alternative representation for B_H is

$$B_H(t) = C_H \int_{-\infty}^{t} \left(\varphi_H(t - u) - \varphi_H(-u) \right) dB(u), \tag{4.3.2}$$

where C_H is a normalizing constant and $\varphi_H(x) = 1_{[0,\infty)}(x)x^{H-1/2}, x \in \mathbb{R}$. Regularization replaces φ_H in (4.3.2) with a new, "better behaved" function φ_H^R so that the regularized process B_H^R has desirable properties like: B_H^R is a Gaussian semimartingale with the same long-range dependence as fractional Brownian motion. This approach can also be applied to $\Delta_{H,\alpha}$, say, and may offer an alternative modeling tool for SDEs driven by such processes. The basic ideas underlying regularization are discussed in [Rog97]; further details and refinements are to be found in [Che00b].

4.4 SELECTED TOPICS ON FRACTIONAL BROWNIAN MOTION

There are numerous results to be found in the literature which treat special properties of fractional Brownian motion (as can easily be found out by doing a literature search on MathSciNet for instance). Below we present some of the results available. We definitely do not strive for completeness but aim more at a sample of interesting properties.

4.4.1 Distribution of the Maximum of Fractional Brownian Motion

In many applications, the study of the supremum (up to a time t) of stochastic processes is important. This problem has been well studied for Brownian motion, a natural question concerns the generalization to other selfsimilar processes.

We consider here fractional Brownian motion $\{B_H(t), t \geq 0\}$. As we have seen, sample paths of $B_H(t)$ with $H > 1/2$ are smoother than those of Brownian motion. For this reason, we only consider the case $H > 1/2$.

Introduce the random variable

$$\xi_T = \max_{0 \leq t \leq T} B_H(t).$$

We are interested in the behavior of the distribution of ξ_T for large T. Note that, by the selfsimilarity of fractional Brownian motion, we have

$$P\left\{ \max_{0 \leq t \leq T} B_H(t) \leq x \right\} = P\left\{ \max_{0 \leq t \leq 1} B_H(t) \leq \frac{x}{T^H} \right\}.$$

Therefore, the problem of establishing the asymptotics of $P\{\max_{0 \leq t \leq T} B_H(t) \leq x\}$ with x fixed and $T \to \infty$ is equivalent to that with fixed T and $x \to 0$.

For Brownian motion ($H = 1/2$), it is well known that for any $x > 0$,

$$P\{\xi_T \leq x\} \sim \text{const } T^{-1/2}. \tag{4.4.1}$$

One wants to generalize (4.4.1) to fractional Brownian motion. The following argument about fractional Brownian motion is due to Sinai [Sin97].

Introduce a random variable τ_x as the first crossing time of level x by $B_H(t)$. Since

$$P\{\xi_T < x\} = P\{\tau_x > T\},$$

it is enough to examine the distribution of τ_x for large T. One can show that the distribution of τ_x has a density p_{τ_x}, say. Sinai [Sin97] moreover showed that p_{τ_x} satisfies a certain Volterra-type integral equation and proved the following result.

Theorem 4.4.1 *For H ($> 1/2$) sufficiently close to $1/2$,*

$$p_{\tau_x}(T) \le \text{const } (2H - 1)T^{H-2+b(H)},$$

for large T, where $|b(H)| \le \text{const } (2H - 1)$.

When we are interested in extremes of processes, we have to investigate the tail probabilities $P\{\xi_T > x\}$ for large x and T fixed. For this problem, there is some literature that treats not only fractional Brownian motion, but general selfsimilar processes with stationary increments. Interested readers should consult [Alb98].

4.4.2 Occupation Time of Fractional Brownian Motion

Define $\mathbb{R}^d$-valued fractional Brownian motion $\{\mathbb{B}_H(t), t \in \mathbb{R}\}$ by

$$\mathbb{B}_H(t) = \left(B_H^{(1)}(t), ..., B_H^{(d)}(t)\right)'$$

where $\{B_H^{(j)}(t)\}$, $1 \le j \le d$, are independent copies of real-valued fractional Brownian motion. For a general discussion of selfsimilar processes on $\mathbb{R}^d$, see Section 9.1.

Let f be a bounded integrable function on $\mathbb{R}^d$ such that $\bar{f} := \int_{\mathbb{R}^d} f(x)\, dx \ne 0$. Let $L_t(x)$, $t \ge 0$, $x \in \mathbb{R}^d$, be a jointly continuous local time of $\{\mathbb{B}_H(t)\}$ defined by

$$\int_0^t g(\mathbb{B}_H(u))\, du = \int_{\mathbb{R}^d} g(x) L_t(x)\, dx$$

for every bounded continuous function g on $\mathbb{R}^d$.

Then it is easily seen by the existence of the local time and the selfsimilarity of fractional Brownian motion that if $0 < Hd < 1$, then

$$\frac{1}{\lambda^{1-Hd}} \int_0^{\lambda t} f(\mathbb{B}_H(s))\, ds \xrightarrow{w} \bar{f} L_t(0)$$

as $\lambda \to \infty$, where $\overset{w}{\to}$ denotes weak convergence over the space $C[0, \infty)$, [KasKos97]. In the critical case where $Hd = 1$, the following result is known.

Theorem 4.4.2 [KasKos97]   *Let $d \geq 2$ and $Hd = 1$. Then*

$$\frac{1}{\lambda} \int_0^{e^{\lambda t}} f(\mathbb{B}_H(s)) \, ds \overset{d}{\Rightarrow} \frac{1}{(\sqrt{2\pi})^d} \bar{f} Z(t),$$

as $\lambda \to \infty$, where $\{Z(t)\}$ is given by

$$P\{Z(t_1) \geq x_1, ..., Z(t_n) \geq x_n\} = \exp\left\{-\frac{x_1}{t_1} - \frac{x_2 - x_1}{t_2} - \cdots - \frac{x_n - x_{n-1}}{t_n}\right\}.$$

(The weak convergence at $t = 1$ was first proved by [Kon96].)

The following result is also known.

Theorem 4.4.3 [KasOga99]   *Suppose $d \geq 2$ and $0 < H < 1$. Let $\alpha = 1/(1 - Hd)$ and*

$$Z_H(t) := \left(\sqrt{2\pi}\right)^d (1 - Hd) L_{t^\alpha}(0).$$

Then as $H \uparrow 1/d$,

$$Z_H(t) \overset{d}{\Rightarrow} Z(t),$$

where $\{Z(t)\}$ is the same as in Theorem 4.4.2.

4.4.3 Multiple Points of Trajectories of Fractional Brownian Motion

Several properties of trajectories of multidimensional fractional Brownian motion with multiparameters have also been studied. Let $\{B_H(t), t \in \mathbb{R}^N\}$ be a mean-zero Gaussian process with covariance

$$E[B_H(t)B_H(s)] = \frac{1}{2}\left\{|t|^{2H} + |s|^{2H} - |t - s|^{2H}\right\}.$$

Consider independent copies $\{B_H^{(j)}(t)\}, j = 1, ..., d$, of $\{B_H(t)\}$ and the process $\{\mathbb{B}_H(t) = (B_H^{(1)}(t), ..., B_H^{(d)}(t))'\}$. This is an $\mathbb{R}^d$-valued fractional Brownian motion with multiparameter $t \in \mathbb{R}^N$. For the Hausdorff measure properties and multiple point properties of the trajectories of $\{\mathbb{B}_H(t)\}$, see [Tal95, Tal98], [Xia97, Xia98] and the references therein. As an example, we present some results from [Tal98]. Let $N = 1$. We consider the problem whether the trajectories of $\{\mathbb{B}_H(t)\}$ have k-multiple points, a problem well studied for Brownian motion.

Theorem 4.4.4 [Tal98]

(i) *If $Hd \geq k/(k-1)$, then the trajectories of $\{\mathbb{B}_H(t)\}$ do not have k-multiple points almost surely.*

(ii) *If $1 < Hd < k/(k-1)$, then $\{\mathbb{B}_H(t)\}$ has k-multiple points almost surely, and the set of k-multiple points of the trajectories is a countable union of sets of finite Hausdorff measure associated with the function $\varphi(\varepsilon) = \varepsilon^{k/H-(k-1)d}(\log\log(1/\varepsilon))^k$.*

(iii) *If $Hd = 1$, then $\{\mathbb{B}_H(t)\}$ has k-multiple points almost surely, and the set of such points is a countable union of sets of finite Hausdorff measure associated with the function $\varphi(\varepsilon) = \varepsilon^d(\log(1/\varepsilon))\log\log\log(1/\varepsilon))^k$.*

4.4.4 Large Increments of Fractional Brownian Motion

The results in this section are due to [ElN99].

Let a_T be a nonincreasing function of $T \geq 0$ such that

$$0 \leq a_T \leq T,$$

$$\frac{a_T}{T} \text{ is a nonincreasing function of } T \geq 0,$$

$$\lim_{T\to\infty} \frac{\ln T/a_T}{\ln_2 T} = r \in [0, \infty].$$

Here we set $\ln u = \log(u \wedge e)$ and $\ln_2 u = \ln(\ln u)$ for $u \geq 0$. Define V_T by

$$V_T = \sup_{0 \leq s \leq T - a_T} \beta_T |B_H(s + a_T) - B_H(s)|,$$

where

$$\beta_T^{-1} = 2^{1/2} a_T^H \left(\ln \frac{T}{a_T} + \ln_2 T\right)^{1/2}.$$

When $H = 1/2$, the behavior of V_T was studied in [CsoRev79, CsoRev81] and [BooSho78]. Their results are stated as follows.

Theorem 4.4.5 *When $H = 1/2$, we have with probability one,*

$$\limsup_{T\to\infty} V_T = 1$$

and

$$\liminf_{T\to\infty} V_T = \sqrt{\frac{r}{r+1}},$$

where $\sqrt{r/(r+1)} = 1$ if $r = \infty$.

When $H \neq 1/2$, general results for V_T were obtained by [Ort89]. He established the following theorem.

Theorem 4.4.6 *We have with probability one,*

$$\limsup_{T \to \infty} V_T = 1,$$

and if $r = \infty$, then

$$\lim_{T \to \infty} V_T = 1.$$

When $H \neq 1/2$ and $r < \infty$, [ElN99] obtained the following.

Theorem 4.4.7 *Set*

$$\tau = \lim_{T \to \infty} \frac{a_T}{T} \in [0, 1].$$

(i) *If $0 < H < 1$ and $\tau > 0$, then we have with probability one,*

$$\liminf_{T \to \infty} V_T = 0.$$

(ii) *If $0 < H \leq 1/2$ and $\tau = 0$, then we have with probability one,*

$$\liminf_{T \to \infty} V_T = \sqrt{\frac{r}{r+1}}.$$

(iii) *If $1/2 < H < 1$, $\tau = 0$ and $r > 4^H/(4 - 4^H)$, then we have with probability one,*

$$\liminf_{T \to \infty} V_T = \sqrt{\frac{r}{r+1}}.$$

An interesting paper considering so-called fast sets and points for fractional Brownian motion is [KhoShi00].

Chapter Five

Selfsimilar Processes with Independent Increments

In this chapter, we discuss selfsimilar processes with independent increments but not necessarily having stationary increments. We call $\{X(t), t \geq 0\}$ which is H-selfsimilar with independent increments as H-ss, ii. Selfsimilar processes discussed in this chapter are $\mathbb{R}^d$-valued, $d \geq 1$.

5.1 K. SATO'S THEOREM

As already seen in Theorem 1.4.2, if selfsimilar processes have independent and stationary increments, then their distributions are stable. Hence, the class of their marginal distributions is determined. However, this is no longer the case for selfsimilar processes without independent and stationary increments. Actually, as mentioned in [BarPer99], there is no simple characterization of the possible families of marginal distributions of selfsimilar processes with only stationary increments. Several authors have looked at this problem. For instance, O'Brien and Vervaat [OBrVer83] studied the concentration function of $\log X(1)^+$ and the support of $X(1)$ in $\mathbb{R}^1$, gave some lower bounds for the tails of the distribution of $X(1)$ in the case $H > 1$, and showed that $X(1)$ cannot have atoms except in certain trivial cases. Also Maejima [Mae86] studied the relationship between the existence of moments and exponents of selfsimilarity, as mentioned in Theorem 3.1.1. One of the interesting questions is: Is the distribution of $X(1)$ outside 0 absolutely continuous if $H \neq 1$? This question was raised by O'Brien and Vervaat [OBrVer83], but as far as we know it is still open.

For selfsimilar processes with independent increments, the situation is better. Below, we use the words marginal and joint in the following way. For an $\mathbb{R}^d$-valued process $\{X(t)\}$, a marginal distribution of $\{X(t)\}$ is the distribution of $X(t)$ on $\mathbb{R}^d$ for any t; for any n, an n-tuple joint distribution of $\{X(t)\}$ is the distribution of $(X(t_1), ..., X(t_n))$ on $\mathbb{R}^{nd}$ for any choice of distinct $t_1, ..., t_n$.

To state the main theorem in this chapter (due to Sato [Sat91]), we start with the notion of *selfdecomposability*.

Definition 5.1.1 *A probability distribution μ on $\mathbb{R}^d$ is called "selfdecomposable" if for any $b \in (0, 1)$, there exists a probability distribution ρ_b such that*

$$\hat{\mu}(\theta) = \hat{\mu}(b\theta)\hat{\rho}_b(\theta), \qquad \forall \theta \in \mathbb{R}^d. \tag{5.1.1}$$

Remark 5.1.1 Selfdecomposable distributions are infinitely divisible.

Property 5.1.1 *Suppose that there exist a sequence of independent $\mathbb{R}^d$-valued random variables $\{X_j\}$, sequences $a_n > 0$, $\uparrow \infty$ and $b_n \in \mathbb{R}^d$ such that for some distribution μ on $\mathbb{R}^d$,*

$$\mathcal{L}\left(a_n^{-1} \sum_{j=1}^n X_j + b_n\right) \to \mu,$$

where $\to$ denotes weak convergence of measures and for every $\varepsilon > 0$, the following asymptotic negligibility condition holds:

$$\lim_{n \to \infty} P\left\{\max_{1 \le j \le n} \left|a_n^{-1} X_j\right| > \varepsilon\right\} = 0.$$

Then μ is selfdecomposable. Conversely, any selfdecomposable distribution can be obtained as such a limit.

Many distributions are known to be selfdecomposable, and their importance has been increasing, for instance, in mathematical finance, turbulence theory and other fields; see, e.g. [Bar98, Jur97].
     The following result links selfsimilarity to selfdecomposability.

Theorem 5.1.1 [Sat91] *If $\{X(t), t \ge 0\}$ is H-ss, ii, then for each t, $\mathcal{L}(X(t))$ is selfdecomposable.*

Proof. Let μ_t and $\mu_{s,t}$ be the distributions of $X(t)$ and $X(t) - X(s)$, respectively. By H-ss, we have that

$$\hat{\mu}_{at}(\theta) = \hat{\mu}_t\left(a^H \theta\right)$$

for any $a > 0$. We also have, for any $b \in (0, 1)$, that

$$\hat{\mu}_t(\theta) = \hat{\mu}_{bt}(\theta)\hat{\mu}_{bt,t}(\theta) = \hat{\mu}_t\left(b^H \theta\right)\hat{\mu}_{bt,t}(\theta).$$

This shows that μ_t is selfdecomposable. $\square$

Sato [Sat91] also showed that for any given $H > 0$ and a selfdecomposable distribution μ on $\mathbb{R}^d$, there exists a uniquely in law H-ss, ii process $\{X(t)\}$ increments such that $\mathcal{L}(X(1)) = \mu$.
     We now know that the one-dimensional (in time) marginal distributions of

H-ss, ii processes are selfdecomposable. What can be said about their n-tuple joint distributions?

Denote the class of all selfdecomposable distributions on $\mathbb{R}^d$ by $L_0(\mathbb{R}^d)$ and write $I(\mathbb{R}^d)$ for all infinitely divisible distributions on $\mathbb{R}^d$. The following sequence of subclasses $L_m(\mathbb{R}^d)$, $m = 0, 1, ..., \infty$, was introduced by Urbanik [Urb72, Urb73] and further studied by Sato [Sat80]. Let m be a positive integer. A distribution μ on $\mathbb{R}^d$ belongs to $L_m(\mathbb{R}^d)$ if and only if $\mu \in L_0(\mathbb{R}^d)$ and, for every $b \in (0, 1)$, ρ_b in (5.1.1) belongs to $L_{m-1}(\mathbb{R}^d)$. The class $L_\infty(\mathbb{R}^d)$ is defined by $\bigcap_{m \geq 0} L_m(\mathbb{R}^d)$. Then we have

$$I(\mathbb{R}^d) \supset L_0(\mathbb{R}^d) \supset L_1(\mathbb{R}^d) \supset \cdots \supset L_\infty(\mathbb{R}^d) \supset S(\mathbb{R}^d),$$

where $S(\mathbb{R}^d)$ is the class of all stable distributions on $\mathbb{R}^d$. A necessary and sufficient condition for distributions to be in $L_m(\mathbb{R}^d)$ in terms of Lévy measures is given in [Sat80].

As shown in Theorem 5.1.1, if $\{X(t), t \geq 0\}$ is a stochastically continuous selfsimilar process with independent increments on $\mathbb{R}^d$, then its marginal distributions are selfdecomposable. However, the n-tuple joint distribution for $n \geq 2$ is not always selfdecomposable; see [Sat91, Proposition 4.2]. In the following theorem we give conditions for joint distributions to be self-decomposable, and furthermore, conditions for them to belong to the smaller classes $L_m(\mathbb{R}^d)$.

Theorem 5.1.2 [MaeSatWat00] *Let $\{X(t), t \geq 0\}$ be a stochastically continuous H-selfsimilar process with independent increments, $H > 0$. Let m be a positive integer or ∞. Then the following four conditions are equivalent. We understand that $m - 1 = \infty$ if $m = \infty$.*

(i) $\mathcal{L}(X(t)) \in L_m(\mathbb{R}^d), \forall t \geq 0.$

(ii) $\mathcal{L}(X(t_1), ..., X(t_n)) \in L_{m-1}(\mathbb{R}^{nd}), \forall n \geq 2, \forall t_1, ..., t_n \geq 0.$

(iii) $\mathcal{L}(\sum_{k=1}^n c_k X(t_k)) \in L_{m-1}(\mathbb{R}^d), \forall n \geq 2, \forall t_1, ..., t_n \geq 0, \forall c_1, ..., c_n \in \mathbb{R}.$

(iv) $\mathcal{L}(X(t) - X(s)) \in L_{m-1}(\mathbb{R}^d), \forall s, t \geq 0.$

Lemma 5.1.1 *Let $m \in \{0, 1, ..., \infty\}$ and $d_1, ..., d_n$ be positive integers. If $\mu \in L_m(\mathbb{R}^{d_1})$ and if T is a linear transformation from $\mathbb{R}^{d_1}$ to $\mathbb{R}^{d_2}$, then $\mu T^{-1} \in L_m(\mathbb{R}^{d_2})$, where $(\mu T^{-1})(B) = \mu(T^{-1}(B))$. If $\mu_k \in L_m(\mathbb{R}^{d_k})$ for $k = 1, ..., n$, then $\mu_1 \times \cdots \times \mu_n \in L_m(\mathbb{R}^d)$ with $d = d_1 + \cdots + d_n$.*

This lemma is essentially found in [Sat80, Theorem 2.4].

Proof of Theorem 5.1.2. Let $0 \leq t_1 \leq \cdots \leq t_n$. Let $Y_1 = X(t_1)$ and $Y_k = X(t_k) - X(t_{k-1})$ for $k = 2, ..., n$. Then $X(t_k) = Y_1 + \cdots + Y_k$. By Lemma 5.1.1 we see that $\mathcal{L}(X(t_1), ..., X(t_n)) \in L_{m-1}(\mathbb{R}^{nd})$ if and only if

$\mathcal{L}(Y_1, ..., Y_n) \in L_{m-1}(\mathbb{R}^{nd})$. Since $Y_1, ..., Y_n$ are independent, Lemma 5.1.1 shows that $\mathcal{L}((Y_1, ..., Y_n)) \in L_{m-1}(\mathbb{R}^{nd})$ if and only if $\mathcal{L}(Y_k) \in L_{m-1}(\mathbb{R}^d)$ for $k = 1, ..., n$. Hence we see that (ii) and (iv) are equivalent. By Lemma 5.1.1, (ii) implies (iii). Obviously (iii) implies (iv). Hence (iii) is equivalent to (ii) and to (iv).

Let us prove the equivalence of (i) and (iv). Let $0 \le s \le t$. Then

$$\hat{\mu}_t(\theta) = \hat{\mu}_s(\theta)\hat{\mu}_{s,t}(\theta) = \hat{\mu}_t\big((s/t)^H \theta\big)\hat{\mu}_{s,t}(\theta), \qquad (5.1.2)$$

where we have used the independent increment property and selfsimilarity with $a = s/t$. On the other hand, by Theorem 5.1.1, $\mu_t \in L_0(\mathbb{R}^d)$. Thus, for any $b \in (0, 1)$, there exists a probability distribution $\rho_{t,b}$ such that

$$\hat{\mu}_t(\theta) = \hat{\mu}_t(b\theta)\hat{\rho}_{t,b}(\theta), \qquad \forall \theta \in \mathbb{R}^d. \qquad (5.1.3)$$

Since $\hat{\mu}_t(\theta) \ne 0$, it follows from (5.1.2), (5.1.3) and taking $b = (s/t)^H$ that

$$\rho_{t,(s/t)^H} = \mu_{s,t}.$$

Hence $\rho_{t,(s/t)^H} \in L_{m-1}(\mathbb{R}^d)$ if and only if $\mu_{s,t} \in L_{m-1}(\mathbb{R}^d)$, concluding that (i) and (iv) are equivalent. $\square$

Note that if $\mathcal{L}(X(1)) \in L_m(\mathbb{R}^d)$, then (i) is true. This is because, by selfsimilarity, $\mathcal{L}(X(t)) = \mathcal{L}(t^H X(1))$.

5.2 GETOOR'S EXAMPLE

Assume $d \ge 3$ and let $\{\mathbb{B}(t)\}$ be a Brownian motion in $\mathbb{R}^d$ with $\mathbb{B}(0) = 0$. For $t > 0$, define

$$L(t) = \sup \{u > 0 : |\mathbb{B}(u)| \le t\}.$$

Since $|\mathbb{B}(t)| \to \infty$ almost surely as $t \to \infty$ when $d \ge 3$, $L(t)$ is finite almost surely.

Getoor [Get79] showed the following.

Theorem 5.2.1 *Let $d \ge 3$. Then the process $\{L(t)\}$ is stochastically continuous and 2-ss, ii. $\{L(t)\}$ has si if and only if $d = 3$.*

Proof. Selfsimilarity can be easily obtained:

$$L(at) = \sup\{u > 0 : |\mathbb{B}(u)| \le at\}$$

$$= \sup\Big\{u > 0 : a^{-1}|\mathbb{B}(u)| \le t\Big\}$$

$$\stackrel{d}{=} \sup\Big\{u > 0 : \big|\mathbb{B}\big(a^{-2}u\big)\big| \le t\Big\}$$

$$= a^2 L(t).$$

As to the other parts of the proof, see [Get79]. $\square$

5.3 KAWAZU'S EXAMPLE

The following examples of ss, ii processes are due to Kawazu (see [Sat91]).
Let $\{B(t)\}$ be a Brownian motion on $\mathbb{R}$ with $B(0) = 0$. Define

$$M(t) = \inf\left\{u > 0 : B(u) - \min_{s \le u} B(s) \ge t\right\},$$

$$V(t) = -\min_{s \le M(t)} B(s)$$

and

$$N(t) = \inf\{u > 0 : B(u) = -V(t)\}.$$

These processes appear in limit theorems of diffusions in random environments. Then the processes $\{M(t)\}$, $\{V(t)\}$ and $\{N(t)\}$ have independent increments, but none of them have stationary increments. The three processes are however selfsimilar, actually $\{M(t)\}$ is 2-ss, $\{V(t)\}$ is 1-ss and $\{N(t)\}$ is 2-ss, which can be seen as follows.

$$M(at) = \inf\left\{u > 0 : B(u) - \min_{s \le u} B(s) \ge at\right\}$$

$$= \inf\left\{u > 0 : a^{-1}\left(B(u) - \min_{s \le u} B(s)\right) \ge t\right\}$$

$$\overset{d}{=} \inf\left\{u > 0 : \left(B(a^{-2}u) - \min_{s \le u} B(a^{-2}s)\right) \ge t\right\}$$

$$= a^2 M(t),$$

$$V(at) = -\min_{s \le M(at)} B(s) \overset{d}{=} -\min_{s \le a^2 M(t)} B(s) = -\min_{s \le M(t)} B\left(a^2 s\right) \overset{d}{=} aV(t)$$

and

$$N(at) = \inf\{u > 0 : B(u) = -V(at)\}$$

$$\overset{d}{=} \inf\{u > 0 : B(u) = -aV(t)\}$$

$$= \inf\left\{u > 0 : a^{-1}B(u) = -V(t)\right\}$$

$$\stackrel{d}{=} \inf\left\{u > 0 : B(a^{-2}u) = -V(t)\right\}$$

$$= a^2 N(t).$$

The $\mathbb{R}^2$-valued process $\{(V(t), N(t))\}$ also has independent increments, but the $\mathbb{R}^3$-valued process $\{(V(t), N(t), M(t))\}$ does not. As to $\{(V(t), N(t))\}$, we also have that

$$\left\{\begin{pmatrix} V(at) \\ N(at) \end{pmatrix}\right\} \stackrel{d}{=} \left\{\begin{pmatrix} a & 0 \\ 0 & a^2 \end{pmatrix}\begin{pmatrix} V(t) \\ N(t) \end{pmatrix}\right\}. \qquad (5.3.1)$$

In Section 9.1 we shall refer to this as operator selfsimilarity of the $\mathbb{R}^2$-valued process $\{(V(t), N(t))\}$. Note that since $\{V(t)\}$ and $\{N(t)\}$ are not independent, selfsimilarity of each process does not imply operator selfsimilarity of $\{(V(t), N(t))\}$, nevertheless (5.3.1) holds.

5.4 A GAUSSIAN SELFSIMILAR PROCESS WITH INDEPENDENT INCREMENTS

The following process is discussed in [NorValVir99].

Let $\{B_H(t), t \geq 0\}$ be a fractional Brownian motion and $1/2 < H < 1$. Define $\{M(t), t \geq 0\}$ by

$$M(t) = \int_0^t u^{1/2-H}(t - u)^{1/2-H} dB_H(u).$$

Then it is shown in [NorValVir99] that the process $\{M(t)\}$

 (i)   is Gaussian,

 (ii) is $(1 - H)$-ss, and

 (iii) has independent increments (but not stationary increments).

$\{M(t)\}$ turns out to be a martingale. This process is useful for the analysis of the first passage time distributions of fractional Brownian motion with positive linear drift. For details, see [NorValVir99] and [Mic99].

Chapter Six

Sample Path Properties of Selfsimilar Stable Processes
with Stationary Increments

6.1 CLASSIFICATION

When two stochastic processes $\{X(t)\}$ and $\{Y(t)\}$ satisfy $\{X(t)\} \overset{d}{=} \{Y(t)\}$, we say that one is a version of the other. If they are modifications of each other, they are also versions of each other.

Typical sample path properties examined in the literature can be summarized as follows:

Property I: There exists a version with continuous sample paths.

Property II: Property I does not hold, but there is a version whose sample paths are right-continuous and have left limits (i.e. are so-called càdlàg).

Property III: Any version of the process is nowhere bounded, i.e. unbounded on every finite interval.

The processes discussed so far can be classified as follows:

Property I: Brownian motion, fractional Brownian motion, linear fractional stable motion for $1/\alpha < H < 1$.

Property II: Non-Gaussian stable Lévy processes.

Property III: Log-fractional stable motion, linear fractional stable motion for $0 < H < 1/\alpha$.

Property I of linear fractional stable motion for $1/\alpha < H < 1$ can be verified by Lemma 4.1.1 as for the case of fractional Brownian motion.

Proofs are needed to justify the classifications of processes with Property III. They can be based on Theorem 6.1.1 below, which is a consequence of Theorem 10.2.3 in [SamTaq94].

Let a $S\alpha S$ process $\{X(t), t \in \mathbb{R}\}$ be given by

$$X(t) = \int_U f(t, u) dW_m(u),$$

where $(U, \mathcal{U}, m)$ is some σ-finite measure space, $f : \mathbb{R} \times U \to \mathbb{R}$ is a function with the property that for each $t \in \mathbb{R}$, $f(t, \cdot) \in L^\alpha(U, \mathcal{U}, m)$, and W_m is a $S\alpha S$ random measure with control measure m such that $E[\exp\{i\theta W_m(A)\}] = \exp\{-m(A)|\theta|^\alpha\}$, $A \in \mathcal{U}$ (see [SamTaq94, Definition 3.3.1]). We assume that $\{X(t)\}$ is stochastically continuous and take a separable version (see [SamTaq94, Definition 9.2.1]). A kernel $f_0(t, u)$ is a modification of $f(t, u)$ if for all $t \in \mathbb{R}$, $f_0(t, \cdot) = f_0(t, \cdot)$ m-a.e. on U. Then $\{X_0(t)\}$ defined by $X_0(t) = \int_U f_0(t, u) dW_m(u)$ is a version of $\{X(t)\}$.

Theorem 6.1.1 *Let $0 < \alpha < 2$. Suppose there is a countable subset T^* of $\mathbb{R}$ such that for every $a < b$ we have $\int_a^b (\sup_{t \in T^*} |f(t, u)|)^\alpha m(du) = \infty$. Then $\{X(t)\}$ has Property III.*

The fact that log-fractional stable motion and linear fractional stable motion for $0 < H < 1/\alpha$ have Property III follows from Theorem 6.1.1; indeed for every $u \in [a, b]$, $\sup_{t \in T^*} |f(t, u)| = \infty$ with $T^* = \{r$ rational, $a \le r \le b\}$ in either case. (See Examples 10.2.5 and 10.2.6 in [SamTaq94].)

6.2 LOCAL TIME AND NOWHERE DIFFERENTIABILITY

For stochastic processes with continuous sample paths, a natural further question addresses sample path differentiability. In this section, we apply an argument in Berman [Ber69] to prove that for $0 < H < 1/2$ an H-ss, si, $S\alpha S$ process is nowhere differentiable. This section is based on [KonMae91a]. We start with a result on the local time of H-ss, si, $S\alpha S$ processes.

Theorem 6.2.1 *Let $\{X(t), t \in T\}$ be an H-ss, si, $S\alpha S$ process with $0 < H < 1$. Then $\{X(t)\}$ has L^2-local time almost surely.*

Proof. Let $I = [a, b]$, $-\infty < a < b < \infty$, and put

$$\mu_X(A) = \text{Leb}\{t \in I, X(t) \in A\}, \qquad A \in \mathfrak{B}(\mathbb{R}).$$

Note that we have suppressed $\omega \in \Omega$ in the above; Leb denotes Lebesgue measure. Let

$$h(\theta) = \int_{-\infty}^{\infty} e^{i\theta x} d\mu_X(x) = \int_I e^{i\theta X(t)} dt.$$

Since $\{X(t)\}$ is H-ss, si, $S\alpha S$, we have that

$$E\left[|h(\theta)|^2\right] = E\left[\int\int_{I \times I} e^{i\theta(X(t) - X(s))} dt \, ds\right]$$

$$= \int_{I \times I} e^{-c|t-s|^{\alpha H}|\theta|^\alpha} \, dt \, ds, \tag{6.2.1}$$

where c is a positive constant determined by

$$E\left[e^{i\theta X(1)}\right] = e^{-c|\theta|^\alpha}.$$

Hence

$$E\left[\int_{-\infty}^\infty |h(\theta)|^2 d\theta\right] = \int_{I \times I} \int_{-\infty}^\infty e^{-c|u|^\alpha} \, du \, \frac{dt \, ds}{|t-s|^H} < \infty,$$

if $0 < H < 1$. Therefore for almost all $\omega \in \Omega$, $h(\theta, \omega)$ is square integrable so that there exists an L^2-occupation density of the occupation measure $\mu_X(\cdot)$, which is the local time. $\square$

Theorem 6.2.2 *Let* $\{X(t), t \in T\}$ *be an H-ss, si, SαS process with* $0 < H < 1/2$, *and let I be a finite interval. Then* $\{X(t)\}$ *satisfies*

$$\frac{(X_M - X_m)}{\text{Leb}(I)} \geq C|\log(X_M - X_m)|^\delta$$

for some positive constants C and δ, where

$$X_M = \sup_{t \in I} X(t) \quad and \quad X_m = \inf_{t \in I} X(t).$$

Hence if X(t) is continuous, then it is nowhere differentiable, and if it is right continuous, then it is nowhere differentiable from the right.

Proof. Fix $\omega \in \Omega$ such that $h(\theta, \omega)$ is square integrable. By the Fourier inversion formula,

$$\text{Leb}(I) = \mu_X((X_m, X_M)) = \int_{-\infty}^\infty \frac{e^{-i\theta X_M} - e^{-i\theta X_m}}{2\pi i \theta} h(\theta) d\theta.$$

Hence for any $\varepsilon > 0$,

$$\text{Leb}(I)^2 = \frac{1}{4\pi^2}\left[\int_{-\infty}^\infty \left|\frac{e^{-i\theta(X_M - X_m)} - 1}{\theta}\right| \frac{1}{|\theta|^{1/2}(|\log|\theta|| + 1)^{(1+\varepsilon)/2}}\right.$$

$$\left. \times |\theta|^{1/2}(|\log|\theta|| + 1)^{(1+\varepsilon)/2}|h(\theta)| d\theta\right]^2$$

$$\leq \int_{-\infty}^\infty \frac{\left|e^{-i\theta(X_M - X_m)} - 1\right|^2}{|\theta|^3(|\log|\theta|| + 1)^{1+\varepsilon}} d\theta$$

$$\times \int_{-\infty}^\infty |\theta|(|\log|\theta|| + 1)^{1+\varepsilon}|h(\theta)|^2 d\theta$$

$$=: I_1 \times I_2. \tag{6.2.2}$$

By (6.2.1), for $0 < H < 1/2$,

$$E[I_2] = \int_{-\infty}^{\infty} |\theta|(|\log|\theta|| + 1)^{1+\varepsilon} E\Big[|h(\theta)|^2\Big] d\theta$$

$$= \int_{I\times I} \int_{-\infty}^{\infty} |\theta|(|\log|\theta|| + 1)^{1+\varepsilon} e^{-c|t-s|^{\alpha H}|\theta|^{\alpha}} d\theta\, dt\, ds$$

$$= \int_{I\times I} \int_{-\infty}^{\infty} |u|\left(\Big|\log\frac{|t-s|^H}{|u|}\Big| + 1\right)^{1+\varepsilon} e^{-c|u|^{\alpha}} du\, \frac{dt\, ds}{|t-s|^{2H}}$$

$$\leq C \int_{I\times I} (|\log|t-s|| + 1)^{1+\varepsilon} \frac{dt\, ds}{|t-s|^{2H}} < \infty.$$

Thus

$$I_2 < \infty \quad \text{a.s.} \tag{6.2.3}$$

As to I_1,

$$I_1 = \int_{-\infty}^{\infty} \frac{\big|e^{-iu} - 1\big|^2 (X_M - X_m)^2}{|u|^3(|\log|(X_M - X_m)/u|| + 1)^{1+\varepsilon}} du$$

$$\leq C \frac{(X_M - X_m)^2}{|\log(X_M - X_m)|^{\varepsilon}}.$$

We thus have from (6.2.1)–(6.2.3) that for some positive constant C,

$$\frac{(X_M - X_m)}{\text{Leb}(I)} \geq C|\log(X_M - X_m)|^{\varepsilon/2}. \qquad \square$$

Chapter Seven

Simulation of Selfsimilar Processes

7.1 SOME REFERENCES

There are numerous textbooks on simulation. For our purposes, an excellent text is Ripley [Rip87], at the more introductory level Morgan [Mor84] and Ross [Ros91]. See also Rubinstein and Melamed [RubMel98]. For the simulation of α-stable processes, see Janicki and Weron [JanWer94]. The presentation of this chapter is mainly based on Asmussen [Asm99]. The latter also contains an excellent list of references related to the simulation of rare events. A short discussion on the simulation of long-memory processes is to be found in Beran [Ber94]. As so often in this field, Benoit Mandelbrot was involved very early on; see for instance [Man71]. An excellent discussion on various simulation routines for selfsimilar processes is to be found on Murad Taqqu's website: http://math.bu.edu/people/murad, look for the link "Statistical methods for long-range dependence".

7.2 SIMULATION OF STOCHASTIC PROCESSES

The simulation of stochastic processes in general may be rather involved depending on whether we have a process defined in discrete or continuous time, one- or more dimensional, defined explicitly or implicitly through some equation(s) (PDE, SDE, recursive equation, ...). Also an important factor concerns which functional of the process one is interested in: for instance a hitting time, a marginal distribution, a sample path. In all of these, the precise stochastic structure of the process plays an important role: stationary or not, specific dependence structures, regularity of sample paths. It is clear that in this brief introduction, we will not be able to enter into details. We give an introduction to the basic methodologies underlying simulation technology for selfsimilar (and more general) stochastic processes. The reader is referred to the references above for more details. Various basic questions will not be touched upon; for example, the assessment of the quality of the simulation methods presented. In order to answer the latter, a deeper study on the use of an appropriate assessment-functional has to be discussed first.

For the simulation of Gaussian processes in general, and Brownian motion in particular, there exist numerous procedures. Suppose $\{X(t)\}$ is a Gaussian process with covariance function $\gamma(s,t) = \text{Cov}(X(s), X(t))$. We will concentrate on the simulation of a finite-dimensional realization (or skeleton) $x(0), \ldots, x(n)$. One group of methods is based on appropriate matrix decomposition of the covariance matrix

$$\Gamma(n+1) = (\text{Cov}(X(i), X(j)))_{i,j=0,\cdots,n+1}.$$

An often used procedure uses the so-called Cholesky decomposition, see [Asm99]. Suppose we want to simulate $X(n+1)$, based on $X(0), X(1), \ldots, X(n)$. Then proceed as follows:

Step 1: write

$$\Gamma(n+1) = \begin{pmatrix} \Gamma(n) & \gamma(n) \\ \gamma(n)' & \gamma(n+1, n+1) \end{pmatrix},$$

where $\Gamma(n)$ is the covariance matrix of $X(0), \ldots, X(n)$ and $\gamma(n)$ is the $(n+1)$-column vector with kth component $\gamma(n+1, k)$, $k = 0, \ldots, n$.

Step 2: the conditional distribution of $X(n+1)$, given $X(0), \ldots, X(n)$ is $N(\widehat{X}(n+1), \sigma_n^2)$, where

$$\widehat{X}(n+1) = \gamma(n)'\Gamma(n)^{-1}\begin{pmatrix} X(0) \\ X(1) \\ \vdots \\ X(n) \end{pmatrix}$$

$$\sigma_n^2 = \gamma(n+1, n+1) - \gamma(n)'\Gamma(n)^{-1}\gamma(n)$$

and generate $X(n+1)$ according to $N(\widehat{X}(n+1), \sigma_n^2)$.

Remark 7.2.1 The Cholesky decomposition enters for the efficient (recursive) calculation of $\gamma(n)'\Gamma(n)^{-1}$. A detailed discussion of this procedure, in the stationary case where $\gamma(i,j) = \gamma(|i-j|)$, is to be found in [Ber94, p. 215]. The general case is discussed in [Asm99, Chapter VIII, 4].

There are various special cases where alternative methods can be used, so for instance for Brownian motion $\{B(t)\}$ itself. Say, we want to simulate $\{B(t)\}$ on a discrete skeleton $0, h, 2h, \ldots$. Then we just generate the increments

$$B(h) = B(h) - B(0), \ B(2h) - B(h), \ B(3h) - B(2h), \ldots$$

as i.i.d. $N(0, 1)$ random variables. Linear interpolation can then be used to approximate a continuous time sample path. Alternatively, we may base simulation on a functional Central Limit Theorem or one of the many series representations of Brownian motion. As mentioned in [Asm99], "In view of the simplicity of this (i.e. the above discrete skeleton) procedure, there is not much literature on the simulation of Brownian motion. A notable exception is [Knu84]." This lack of literature is unfortunate, as one very quickly enters into simulation questions regarding $\{B(t)\}$ which are not so trivial. See [Asm99] for examples involving the calculation of hitting times and the simulation of reflected Brownian motion: these questions are especially important in insurance (ruin, say) and finance (e.g. barrier options).

Also note that for a wide class of models (especially in econometrics) one has explicit recursive equations from which simulation becomes fairly straightforward. For example, for the class of ARMA(p, q) processes,

$$X(t) - \alpha_1 X(t-1) - \cdots - \alpha_p X(t-p) = \varepsilon_t + \theta_1 \varepsilon_{t-1} + \cdots + \theta_q \varepsilon_{t-q},$$

where the $\{\varepsilon_k\}$ are i.i.d. $N(0, \sigma^2)$. Similarly, for examples like the ARCH(1) process,

$$\begin{cases} X(t) = \sigma_t \varepsilon_t \\ \sigma_t^2 = \lambda + \beta X(t-1)^2 \end{cases}$$

and various generalizations.

7.3 SIMULATING LÉVY JUMP PROCESSES

We know that the distributions of the increments of a Lévy process are infinitely divisible. Hence a basic step in the simulation of Lévy processes is the simulation from general infinitely divisible distributions. For this, see Bondesson [Bon82]. The special properties of the Lévy measure in the Lévy–Kolmogorov representation plays a crucial role here. Of course, in general we may write a Lévy process $\{X(t)\}$ as

$$X(t) = \mu t + \sigma B(t) + J(t)$$

that is, the independent sum of a linear drift, a Brownian component and a jump process determined in terms of the Lévy measure ν of the process. The process $\{J(t)\}$ can again be decomposed in $J(t) = J^{(1)}(t) + J^{(2)}(t)$ where $\{J^{(2)}(t)\}$ is a compound Poisson process for which simulation is trivial ($J^{(1)}$ and $J^{(2)}$ are independent processes). For $J^{(1)}$ various methods exist ranging from neglecting it (after a careful choice of a "cut-off" point leading to the decomposition $J = J^{(1)} + J^{(2)}$ in the first place) to replacing it by a suitably chosen Brownian motion. For details on this, see [Asm99].

Of course, using [Bon82] and the fact that a Lévy process has stationary, independent increments, one can in several cases fairly easily simulate a discrete skeleton of $\{X(t)\}$, a jump Lévy process. Examples include gamma, Cauchy and inverse Gaussian processes. The special case of an α-stable Lévy process has generated a lot of interest in the literature. The standard algorithm for the generation of α-stable distributions is due to Chambers, Mallows and Stuck [ChaMalStu76]. See also [SamTaq90] and [JanWer94]. The latter reference contains a detailed discussion on the simulation of α-stable processes;

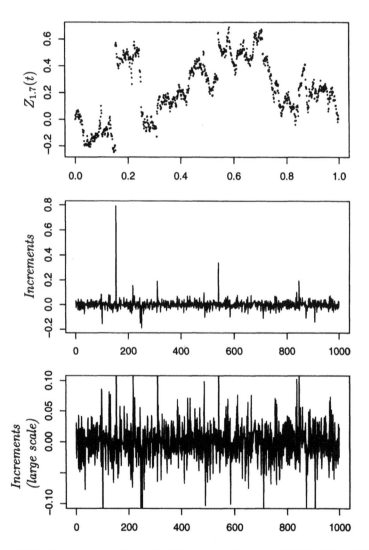

Figure 7.3.1 Sample paths of α-stable Lévy motion for $\alpha = 1.7$ ($H = 1/\alpha = 0.59$), with the corresponding increments (below on a larger scale). The paths are not continuous, and therefore they are displayed as a set of points.

see also [Asm99, Chapter VIII, 2] and references therein. The key point is that, due to the infinity of jumps, some truncation or limiting procedure is called for. Asmussen [Asm99] also discusses an approach based on a series representation. Finally see http://academic2.american.edu/~jpnolan/stable/ stable.html, the webpage of John Nolan, for interesting information and software on stable distributions.

As an example (Figure 7.3.1) we have simulated a realization of an α-stable Lévy process $\{Z_\alpha(t)\}$ with $\alpha = 1.7$.

7.4 SIMULATING FRACTIONAL BROWNIAN MOTION

Recall from Definition 1.3.1 that $\{B_H(t), 0 < H \le 1\}$, is a fractional Brownian motion if it is a mean zero Gaussian process with covariance function.

$$\gamma(s,t) = \frac{1}{2}\left\{t^{2H} + s^{2H} - |t-s|^{2H}\right\}E\left[B_H(1)^2\right].$$

For $H = 1/2, \{B_{1/2}(t)\}$ is a Brownian motion. Of course, one can use the Cholesky decomposition-based method from Section 7.2. For an implementation of the latter method, see Michna [Mic98a, Mic98b, Mic99]. For an implementation based on the Fast Fourier Transform (FFT), see [Ber94]. The latter reference also contains S-Plus code for the simulation of fractional Brownian motion and fractional ARIMA processes. Asmussen [Asm99] warns of the use of the FFT and so-called ARMA approximations as they may destroy the long-range dependence. One could use representation theorems like Theorem 1.3.3, but also here, truncation will destroy long-range dependence. An alternative representation for which truncation is not needed, is given in [NorValVir99]; see also [Asm99], where further interesting references (including work on importance sampling) can be found. Finally, wavelets have also entered the fractional Brownian motion scene; see for example [Whi01] and the references therein.

In Figures 7.4.1–7.4.5 we have simulated sample paths of fractional Brownian motion with $H = 0.1, 0.3, 0.5, 0.7, 0.9$. The increasing smoothness of the sample paths as explained in Section 4.1 is clearly visible. We also plotted the autocorrelation function, see Section 3.2 for more details.

The following links show how to generate paths of fractional Brownian motion in Mathematica (a free Mathematica Reader is available from http://www.wolfram.com):
http://didaktik.phy.uni-bayreuth.de/mathematica/meader_2/htmls/2-08.htm,
http://www.mathconsult.ch/showroom/pubs/MathProg/htmls/2-08.htm.
For an excellent discussion, see also [SamTaq94, Section 7.11].

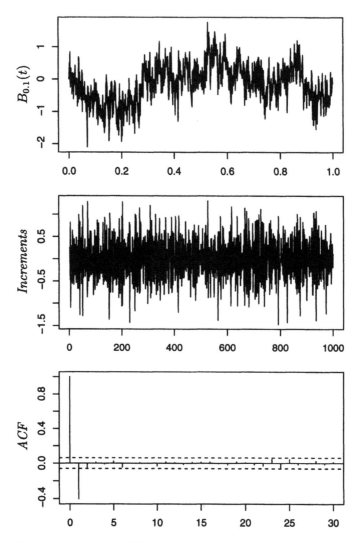

Figure 7.4.1 Sample paths of fractional Brownian motion for $H = 0.1$, with the corresponding increment process and sample autocorrelation function. The correlations decay fast and are negative, as expected.

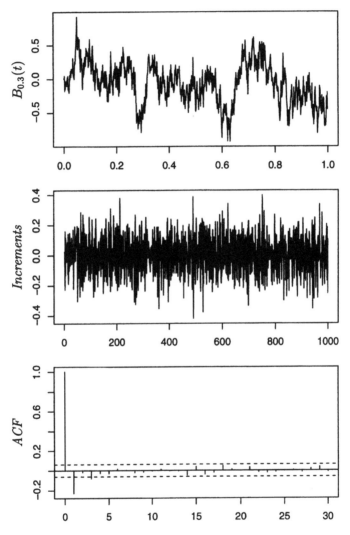

Figure 7.4.2 Sample paths of fractional Brownian motion for $H = 0.3$, with the corresponding increment process and sample autocorrelation function. The correlations decay fast and are negative, as expected.

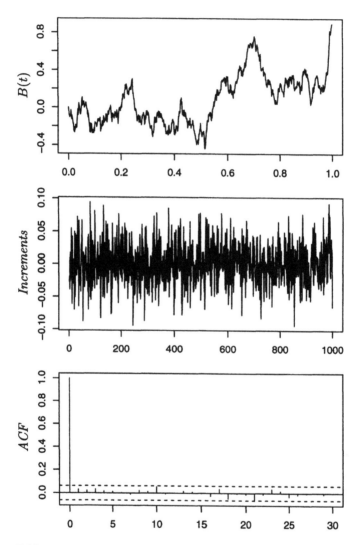

Figure 7.4.3 Sample paths of Brownian motion ($H = 0.5$), with the corresponding increment process and sample autocorrelation function. As expected, the correlations are negligible.

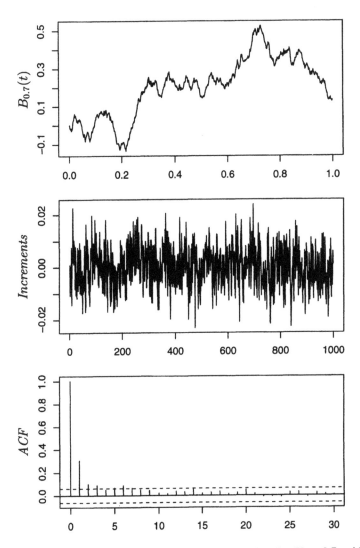

Figure 7.4.4 Sample paths of fractional Brownian motion for $H = 0.7$, with the corresponding increment process and sample autocorrelation function. The correlations are positive and decay slower than in the case $H = 0.1$ displayed in Figure 7.4.1. The process exhibits long-range dependence.

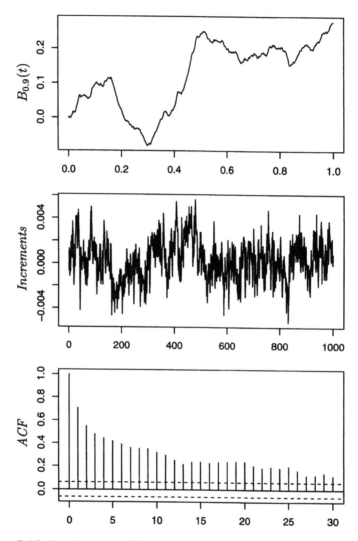

Figure 7.4.5 Sample paths of fractional Brownian motion for $H = 0.9$, with the corresponding increment process and sample autocorrelation function. The correlations clearly decay much more slowly than in the case $H = 0.1$ displayed in Figure 7.4.1: the process exhibits long-range dependence.

7.5 SIMULATING GENERAL SELFSIMILAR PROCESSES

It should be clear from the above discussion that for selfsimilar processes which do not belong to the classes already discussed, very little specific tools are available. Clearly, for a specific process given, one may well find a workable approach. Some of the previous references mentioned contain such examples. In general, one may also want to look for "easier" stochastic processes exhibiting the required selfsimilar behavior (long-range dependence); an interesting paper in this context is [Dal99]. We will come back to these, more statistical, issues in the next chapter.

For a simulated path of linear fractional stable motion for $\alpha = 1.7$ and $H = 0.7$}, see Figure 7.5.1. Finally, Figure 7.5.2 contains a simulation of linear fractional stable motion for $\alpha = 1.7$ and $H = 0.9$, to be compared with Figure 7.4.5.

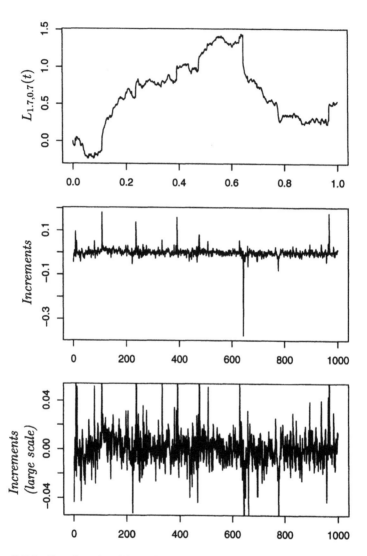

Figure 7.5.1 Sample paths of linear fractional stable motion for $\alpha = 1.7$ and $H = 0.7$, with the corresponding increments. Compared to fractional Gaussian noise with $H = 0.7$ (see Figure 7.4.4), the increments show the expected large values due to the long tail of the distribution.

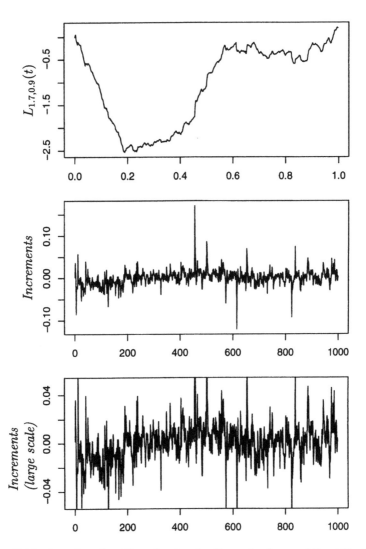

Figure 7.5.2   Sample paths of linear fractional stable motion for $\alpha = 1.7$ and $H = 0.9$, with the corresponding increments. Compared to fractional Gaussian noise with $H = 0.9$ (see Figure 7.4.5), the increments show the expected large values due to the long tail of the distribution.

Chapter Eight

Statistical Estimation

In this chapter we briefly discuss some methods to detect the presence of a long-range dependence structure in a data set. In particular, we present ways to estimate the exponent H. More details and further estimation methods can be found in [Ber94], a text we follow closely. An excellent review of various existing methods, together with examples can be found on Murad Taqqu's website: http://math.bu.edu/people/murad under the heading "Statistical methods for long-range dependence".

8.1 HEURISTIC APPROACHES

The methods described are mainly useful as simple diagnostic tools. Since the results in this section are often difficult to interpret, for statistical inference, more efficient methods exist, such as the maximum likelihood techniques discussed in Section 8.2.

One can come up with various graphical methods as diagnostic tools for selfsimilarity. For instance, suppose $\{X(t)\}$ is H-ss, si with H such that sufficiently high moments of $|X(1)|$ exist; see Theorem 3.1.1. Since $X(t) \overset{d}{=} t^H X(1)$, for any such kth moment, say, we have that

$$m_k(t) = E\left[X(t)^k\right] = t^{kH} E\left[X(1)^k\right],$$

hence (leaving out absolute value signs where necessary)

$$\log m_k(t) = kH \log t + \log E\left[X(1)^k\right].$$

A diagnostic plot therefore could be

$$\{(\log \widehat{m}_k(t), \ \log t), \ t \in T\}$$

over a set T of t-values and for relevant values of k. Here $\widehat{m}_k$ denotes some moment-type estimator. The log-log linearization above can be found for various functionals of ss processes; see for instance Sections 8.1.1–8.1.3 below.

8.1.1 The *R/S*-Statistic

This method was first proposed by Hurst [Hur51], and is based on the following definition.

Definition 8.1.1 *Let* $\{X_i\}$ *be a sequence of random variables, and* $Y_n = \sum_{i=1}^n X_i$. *Define the "adjusted range"*

$$R(\ell, k) = \max_{1 \le i \le k} \left\{ Y_{\ell+i} - Y_\ell - \frac{i}{k}(Y_{\ell+k} - Y_\ell) \right\}$$

$$- \min_{1 \le i \le k} \left\{ Y_{\ell+i} - Y_\ell - \frac{i}{k}(Y_{\ell+k} - Y_\ell) \right\}$$

and

$$S(\ell, k) = \left\{ \frac{1}{k} \sum_{i=\ell+1}^{\ell+k} (X_i - \overline{X}_{\ell,k})^2 \right\}^{1/2}, \quad \ell \ge 0, \quad k \ge 1,$$

where $\overline{X}_{\ell,k} = k^{-1} \sum_{i=\ell+1}^{\ell+k} X_i$. *The ratio*

$$Q(\ell, k) = \frac{R(\ell, k)}{S(\ell, k)} \tag{8.1.1}$$

is then called the "rescaled adjusted range" or R/S-statistic.

Given a stochastic process $\{X(t)\}$, we denote $Q(k) = Q(0, k)$, the *R/S*-statistic calculated over the first k-block $X(1), ..., X(k)$. The index ℓ in $Q(\ell, k)$ allows us to move this k-block ℓ steps to the right. For a strictly stationary process $X(t)$, one would typically study the behavior of $Q(k) = R(k)/S(k)$, for $k \to \infty$.

  Suppose given a data set $X(1), ..., X(n)$, to estimate the long-memory parameter H, assuming it exists, the logarithm of $Q(k)$ is plotted against $\log k$. For each k, there are $n - k + 1$ replicates $Q(k) = Q(0, k), ..., Q(n - k, k)$. As an illustration, consider the following two examples in Figures 8.1.1 and 8.1.2: (1) A simulated series of fractional Gaussian noise of length $n = 1000$ with parameter $H = 0.9$ (see Figure 7.4.5) and (2) a series of 1000 independent standard normal random variables. The logarithm of Q versus $\log k$ is displayed in Figures 8.1.1 and 8.1.2, respectively, for $k = 10d$ ($d = 1, ..., 20$) and $\ell = 100m$ ($m = 0, 1, 2, ...$).

  The plots are then to be interpreted in the sense of the following theorems (see [Man75] and also [Ber94, pp. 81–82]):

Theorem 8.1.1 *Let* $\{X(t)\}$ *be a strictly stationary stochastic process such that* $\{X(t)^2\}$ *is ergodic and* $n^{-1/2} \sum_{j=1}^{[nt]} X(j)$ *converges weakly to Brownian motion as n tends to infinity. Then, as* $k \to \infty$,

$$k^{-1/2} Q(k) \xrightarrow{d} \xi,$$

where ξ *is a nondegenerate random variable.*

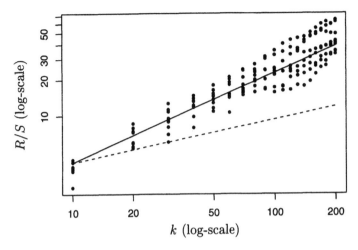

Figure 8.1.1 *R/S* plot for simulated fractional Brownian motion with $H = 0.9$. The least squares fit is based on the data for k in the range 50–200. The line has slope 0.884. A dashed line with slope 0.5 is included for reference.

The assumptions of Theorem 8.1.1 hold for most common short-memory processes. One may say that whenever the central limit theorem holds, the statistic $k^{-1/2}Q(k)$ converges to a nondegenerate random variable and hence we expect a situation like that shown in Figure 8.1.2 to occur.

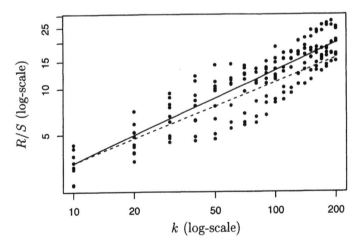

Figure 8.1.2 *R/S* plot for simulated Brownian motion. The line has slope 0.553. A dashed line with slope 0.5 is included for reference.

Theorem 8.1.2 *Let* $\{X(t)\}$ *be a strictly stationary stochastic process such that* $\{X(t)^2\}$ *is ergodic and* $n^{-H} \sum_{j=1}^{[nt]} X(j)$ *converges weakly to fractional Brownian motion as n tends to infinity. Then, as* $k \to \infty$,

$$k^{-H} Q(k) \xrightarrow{d} \xi,$$

where ξ *is a nondegenerate random variable.*

For statistical applications, this means that in the plot of $\log Q$ against $\log k$, the points should ultimately (for large values of k) be scattered randomly around a straight line with slope $1/2$ in the former case, and $H > 1/2$ in the case where long-memory exists (using the approximation $\log Q \xrightarrow{d} \log \xi + H \log k$ for k large).

An interesting, so-called robustness property of the R/S statistic is that the asymptotic behavior in Theorem 8.1.1 remains unaffected by long-tailed marginal distributions, in the following sense [Man75]:

Theorem 8.1.3 *Let* $\{X(t)\}$ *be i.i.d. random variables with* $E[X(t)^2] = \infty$, *in the domain of attraction of a stable distribution with index* $0 < \alpha < 2$. *Then the conclusion of Theorem 8.1.1 holds, that is*

$$k^{-1/2} Q(k) \xrightarrow{d} \xi,$$

where ξ *is a nondegenerate random variable.*

Thus, even if $\{X(t)\}$ has a long-tailed marginal distribution, the R/S statistic still reflects the independence in that the asymptotic slope in the R/S plot remains $1/2$.

In the proofs of the above and similar theorems one has to be careful with respect to the precise topologies used. An excellent paper treating these results in detail is [AvrTaq00]. The following proposition quoted from the latter paper and essentially due to Mandelbrot [Man75] is the key result underlying Theorems 8.1.1–8.1.3.

Proposition 8.1.1 *Let* $\{X(t)\}$ *be a strictly stationary sequence such that*

$$\left(\frac{1}{n^{H_1} L_1(n)} \sum_{i=1}^{[nt]} X(i), \ \frac{1}{n^{H_2} L_2(n)} \sum_{i=1}^{[nt]} X(i)^2 \right) \xrightarrow{*} (U(t), \ V(t)),$$

where L_1 *and* L_2 *are slowly varying functions and where* $\xrightarrow{*}$ *denotes convergence in some functional sense, strong enough to imply convergence of the* $\sup_{0 \le t \le T}$ *and* $\inf_{0 \le t \le T}$ *functionals. Then the properly normalized R/S statistic process*

$$\left\{ \frac{R([kt])}{k^J L(k) S(k)}, \ 0 \le t \le 1 \right\} \tag{8.1.2}$$

converges weakly as $k \to \infty$, where L is slowly varying and

$$J = H_1 - H_2/2 + 1/2.$$

It is further discussed in [AvrTaq00] that the objective of an *R/S* analysis of time series is indeed to determine whether there exists $0 \le J \le 1$ and a slowly varying function L so that the process (8.1.2) converges weakly as $k \to \infty$, and also to estimate J. The latter exponent is called the *R/S* exponent. Avram and Taqqu call *R/S* robust if J depends on the exponent characterizing long-range dependence but does not depend otherwise on the underlying distribution of the stationary time series. Hence *R/S* is robust if $J = d + 1/2$ where $d \in (0, 1/2)$ is a measure of long-range dependence of the time series $\{X(t)\}$. The authors of [AvrTaq00] give many examples to calculate d including FARIMA time series and moving averages attracted to a stable Lévy process.

Based on such results, [Ber94, p. 84] summarizes the *R/S* method as follows:

(i)   Calculate $Q(\ell, k)$ for all possible (or for a sufficient number of different) values of ℓ and k.

(ii) Plot $\log Q(\ell, k)$ against $\log k$ for various values of k, over a range of ℓ-values.

(iii) Draw a straight line $y = a + b \log k$ that corresponds to the ultimate (large k) behavior of the data. Estimate the coefficients a and b (by least squares, for instance), and then set $\hat{H} = \hat{b}$.

From a statistical point of view it must be stressed that there are several drawbacks in using this procedure. First, it is difficult to decide from which k the asymptotic behavior starts, and so how many points are to be included in the least squares regression. For finite samples, the distribution of Q is neither normal nor symmetric, and the values of Q for different time points and lags are not independent from each other. This raises the question of whether least squares regression is appropriate. Moreover, only very few values of Q can be calculated for large values of k, thus making the inference less reliable even at large lags.

Because of these problems, [Ber94} concludes that it seems difficult to evaluate the results of statistical inference based on the *R/S* method.

8.1.2 The Correlogram

The plot of the sample correlations $\hat{\rho}(k)$ against the so called lag k (*correlogram*) is a standard diagnostic tool in time series analysis, see for instance [BroDav91]. If a process has uncorrelated increments, then under fairly general conditions the $\sqrt{n}\widehat{\rho}(k)$ are asymptotically independent standard normal random variables, and a correlation can be considered significant at the 5% level if it exceeds the $\pm 2/\sqrt{n}$ bounds. One has to be careful in interpreting simultaneous confidence bands across a wide range of k-values (lags). Spurious values can occur.

The above fairly straightforward asymptotic confidence bounds may not necessarily apply for processes that have correlated increments, and particularly for processes with long-range dependence. Moreover, long memory depends rather on the slow speed of decay of the correlations, and not on the values of the single correlations, which can be arbitrary small.

However, the plot of $\log|\rho(k)|$ against $\log k$ (rather than $\rho(k)$ against k) can be useful to detect long-range dependence. In fact, as the correlations of a long-memory process decay with a rate proportional to k^{2H-2} (see Section 3.2), then for large lags, the points should be scattered around a straight line with negative slope approximately equal to $2H - 2$. In contrast, for short-memory processes, the log-log correlogram should diverge to minus infinity at a rate that is at least exponential.

Figures 8.1.3 and 8.1.4 display the correlogram and the log-log correlogram for fractional Brownian motion with $H = 0.9$ and the Brownian motion of the previous section (see Figures 8.1.1 and 8.1.2). Essentially the same remarks regarding the difficulties of interpreting the plots as for the R/S method apply here. In particular, the log-log plot is mainly useful if the long-range dependence is strong, or if the series is very long. Otherwise, a reliable conclusion can be hardly taken, as Figure 8.1.4 clearly shows.

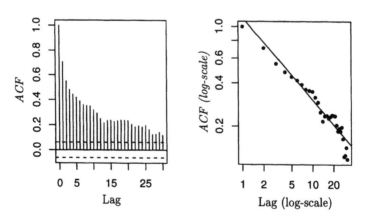

Figure 8.1.3 Correlogram and log-log correlogram for fractional Brownian motion with $H = 0.9$. The slow decay of the sample correlations is clearly visible on the left. The least squares fit in the log-log plot yields a slope of -0.23, that is $\widehat{H} = 0.875$.

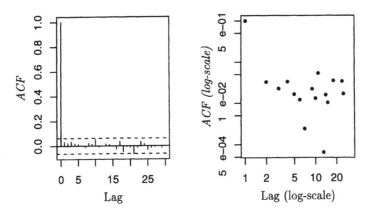

Figure 8.1.4   Correlogram and log-log correlogram for Brownian motion. The sample correlations are negligible, as expected, while the points in the log-log plot do not suggest any conclusion.

8.1.3 Least Squares Regression in the Spectral Domain

The asymptotic behavior at the origin of the spectral density of the increments of an H-ss, si process with $1/2 < H < 1$ is given by

$$f(\lambda) \sim c_f |\lambda|^{1-2H}, \qquad |\lambda| \to 0,$$

for some finite constant c_f; see [Ber94, p. 53] for details. Hence the asymptotic property (for $|\lambda| \to 0$) to be exploited becomes

$$\log f(\lambda) \sim \log c_f + (1 - 2H) \log |\lambda|. \tag{8.1.3}$$

In order to find an estimated version of (8.1.3) on which diagnostic checking for linearity and estimation of c_f and more importantly H can be done, one needs to replace $f(\lambda)$ by some spectral estimator $I_n(\lambda)$ say. Various standard tools from the realm of spectral analysis for time series can now be brought to bear. The details, together with further references, are to be found in [Ber94, Section 4.6].

8.2 MAXIMUM LIKELIHOOD METHODS

In the previous section, we recalled some heuristic methods for estimating the exponent H. These methods may be useful for checking whether a data set exhibits long memory or not; however, we have no corresponding statistical inference theory built on them.

An estimation procedure obtained by building a parametric model and maximizing the corresponding likelihood is often preferable, since it allows one to model the form of the entire correlation structure or, correspondingly, of the spectral density, rather than only describing, and statistically exploiting, its asymptotic properties.

In the previous section, we mentioned that the choice of cut-off point (in either the time domain, k large, or the spectral domain, λ small) was crucial. The methods used were built on asymptotics. In order to perform a full likelihood approach, some extra modeling assumptions (typically for k small and/ or λ large, say) will have to be made. Such assumptions may be available in the data anyhow; see [Ber94, p. 100] for a discussion on this. At the same time, we open the door for model risk, and some criteria for model choice among competing candidates have to be looked at.

In the following we will consider only methods based on Gaussian processes for reasons of simplicity. Various generalizations can be considered.

Beran [Ber94, Sections 5.1–5.2] considers the following set-up. Suppose that $X(1), ..., X(n)$ is a stationary Gaussian sequence with mean zero and variance σ^2, and suppose that the correlations $\rho(k)$ decay at a rate proportional to k^{2H-2}, i.e. the sequence exhibits long-range dependence if $1/2 < H < 1$.

Now consider a family of spectral densities $\{f(\lambda; \theta), \theta \in \Theta \subset \mathbb{R}^M\}$ characterized by the unknown finite dimensional parameter vector

$$\theta = (\sigma^2, H, \theta_3, ..., \theta_M)'.$$

Furthermore assume that $\{X(i), i = 1, ..., n\}$ has a representation of the form

$$X(i) = \sum_{k=0}^{\infty} \psi_k \varepsilon_{i-k}, \qquad (8.2.1)$$

where the ε_k are uncorrelated random variables with zero mean and variance σ_ε^2. The coefficients have to satisfy certain summability conditions in order for $\rho(k)$ to decay as assumed (see [Ber94, Lemma 5.1]).

Clearly, within the model set-up above, one can calculate the exact Gaussian likelihood function and estimate the resulting parameters. Under some extra regularity conditions, it can be shown that the MLE $\hat{\theta}_n$ is strongly consistent, and that

$$\sqrt{n}\left(\hat{\theta}_n - \theta\right) \xrightarrow{d} \xi, \qquad \text{as } n \to \infty, \qquad (8.2.2)$$

where ξ is an M-dimensional random vector with zero mean and covariance matrix, the inverse of a Fisher information matrix calculated via the functional form of the spectral density. It may be somewhat surprising that the $\sqrt{n}$-rate of convergence does not depend on H. Dahlhaus [Dah89] proved that the above estimation procedure is asymptotically efficient.

Clearly, even for the most straightforward models of the type (8.2.1), various computational problems arise. Consequently, as within standard time series analysis, computational shortcuts and approximations to the resulting likelihood function are called for. One particular type of approximation leads to the famous Whittle estimator proposed in [Whi51]; see also [Whi53]. For a discussion in the context of time series models with heavy-tailed innovation, see [EmbKluMik97, Section 7.5]. An excellent discussion in the case of selfsimilar processes is to be found in [Ber94, Section 5.5]. In order to get a feeling for the main ideas, the exact likelihood function is a function of log $|\Sigma(\theta)|$ and $\Sigma(\theta)^{-1}$ where Σ is the covariance matrix of the Gaussian data.

The approximate MLE approach is based on the following (see [Ber94, p. 109]):

(i) $\lim_{n\to\infty} 1/n \log |\Sigma_n(\theta)| = (2\pi)^{-1} \int_{-\pi}^{\pi} \log f(\lambda; \theta) \, d\lambda$ and

(ii) replace $\Sigma(\theta)^{-1}$ by $A(\theta) = (a(j - \ell))_{j,\ell=1,\dots,n}$, where

$$a_{j\ell} = a(j - \ell) = (2\pi)^{-2} \int_{-\pi}^{\pi} \frac{e^{i(j-\ell)\lambda}}{f(\lambda; \theta)} \, d\lambda. \qquad (8.2.3)$$

The Whittle estimator is obtained by minimizing

$$L_W(\theta; x) = \frac{1}{2\pi} \int_{-\pi}^{\pi} \log f(\lambda; \theta) \, d\lambda + \frac{x' A(\theta) x}{n}, \qquad x \in \mathbb{R}^n. \quad (8.2.4)$$

One can show that the resulting (Whittle-)estimator $\hat{\theta}_{n,W}$ is strongly consistent and $\sqrt{n}(\hat{\theta}_{n,W} - \theta) \xrightarrow{d} \xi$ where the random variable ξ is as in (8.2.2).

Thus, Whittle's approximate maximum likelihood estimator has the same asymptotic distribution as the exact MLE, and is therefore asymptotically efficient for Gaussian processes. From the point of view of computation, a consistent improvement can be obtained by replacing the integral (8.2.3) by a sum and then using the Fast Fourier Transform algorithm.

The Whittle approximate maximum likelihood estimator requires the integral (8.2.3) to be evaluated n times for each trial value of θ, thus consuming a large amount of computation time. However, in contrast to the spectral density f itself, the function $1/f$ that appears in (8.2.3) is well behaved at the origin, making possible an approximation of the form

$$\tilde{a}(k) = 2 \frac{1}{(2\pi)^2} \sum_{j=1}^{m} \frac{1}{f\left(\lambda_{j,m}; \theta\right)} e^{ik\lambda_{j,m}} \frac{2\pi}{m},$$

where

$$\lambda_{j,m} = \frac{2\pi j}{m}, \qquad j = 1, 2, \dots, m^*$$

and m^* is the integer part of $(m - 1)/2$. By writing (8.2.4) in terms of the periodogram $I(\lambda; x)$ as

$$L_W(\theta; x) = \frac{1}{2\pi} \left\{ \int_{-\pi}^{\pi} \log f(\lambda; \theta) d\lambda + \int_{-\pi}^{\pi} \frac{I(\lambda; x)}{f(\lambda; \theta)} d\lambda \right\},$$

one obtains the approximation

$$\tilde{L}_W(\theta;x) = 2\frac{1}{(2\pi)^2}\left\{\sum_{j=1}^{m^*}\log f\left(\lambda_{j,m};\theta\right)\frac{2\pi}{m} + \sum_{j=1}^{m^*}\frac{I\left(\lambda_{j,m};x\right)}{f\left(\lambda_{j,m};\theta\right)}\frac{2\pi}{m}\right\}.$$

Because the periodogram can be calculated by the Fast Fourier Transform, $\tilde{L}_W$ can be computed very fast. See [Ber94] for further details and [MikStr01] for an excellent discussion of Whittle estimation in heavy-tailed time series.

At this point, the reader may wonder why we did not (explicitly) use Lamperti's transformation (see Theorem 1.5.1) transforming any selfsimilar process into a stationary one, and then use statistical tools from the realm of stationary processes. See for instance [NuzPoo00] for an estimation approach based on this idea.

8.3 FURTHER TECHNIQUES

The discussion on statistical estimation for selfsimilar processes presented above corresponds to the level of knowledge as reviewed in the book by Beran [Ber94]. The latter contains of course a lot more material. Since its appearance, however, numerous papers have been written on the statistical methodological aspects, as well as on applications of long-range dependence in very diverse fields. An excellent survey, mainly concentrating on an extensive simulation based comparison of various (eight in all) statistical estimation procedures for long-range dependence is [TaqTev98].

On the more methodological side, one of the key issues emerging is the one that no doubt was on the scene from the beginning: "What do we actually mean by long-range dependence?". This issue is still far from resolved. So far, we have concentrated on notions based on the (typically slow) decay of autocorrelations. Various authors have questioned these "definitions". For an example of the sort of alternative approaches being considered, see [ManRacSam99, RacSam01]. These papers also contain various applications within finance and telecommunications (internet traffic). An interesting, thought provoking article on long-range phenomena in large data sets is [Ham87].

An issue which comes up over and over again concerns the determination of long-range dependence when some form of nonstationarity is present. An early paper on this is [BhaGupWay83]. See also [Kue86]. More recently, discussions of this type have emerged from applications in finance, the tenor being that apparent long-range dependence as a statistical artifact stems from other random phenomena within a short-range model. One such result is to be found in the realm of switching (GARCH) regime models where classical tests indicate the presence of long-range dependence, however the models are short-range dependent. See for instance [MikSta00, MikSta01]. This problem is discussed from a more econometric perspective in [DieIno99].

One particular area of quantitative finance where selfsimilarity, long-range dependence and scaling are omnipresent is the field of tick-by-tick (high-density) data analysis. For instance, in the excellent paper [Guiea97], as one of the new stylized facts for intra-daily foreign exchange markets, under Fact 10: Fractal Structure, the authors highlight "Another very striking fact is the regular fractal structure of the FX rates in the sense of Mandelbrot." The scaling law for the volatility is found to be significantly different from the square-root law and seems to be consistently around 0.58 (rather than 0.5). This conclusion was challenged in a recent paper [BarPra01]. These authors claim that the observed scaling phenomenon different from 0.5 is largely due to the semi-heavy tailedness of the distributions concerned rather than to real scaling. This issue will definitely be taken up further. See for instance [BadPol99] for a discussion on scaling in physics. Mandelbrot [Man97] reviews his contributions with applications in finance and risk in mind. Scaling plays a predominant role, as can be seen from the title of [Man97]. Guillaume et al. [Guiea97] contains discussions on short- and long-term memory (see their Fact 9). For a textbook summary of the high frequency finance work, see [Dacetal01].

At this point, we would also like to mention the vast literature on nonlinear time series (see for instance [Ton90]), mixture models (for instance [WonLi00] and the references therein) and fractionally integrated ARMA models, such as [Bai96, LinLi97]. Some interesting early contributions are [Gra80, LiMcL86]. The literature on the above models is huge; the references given should be viewed as a starting point. An interesting book bringing together a lot of material linking heavy tailedness and long memory, mainly from a financial applications point of view is [RacMit00]. There also exist numerous papers on the estimation for long-memory processes using wavelets. The paper of Whitcher [Whi00] and the references therein is a good place to start. See also [Jen99].

Finally, the issue of model deviations concerning dependence is a key issue. How do statistical estimation procedures behave if we have (possibly only small) changes in the dependence structure of the data or model at hand. In general, when modeling dependence, short- and long-range dependence cannot be told apart from data alone, hence model selection, and consequently model risk will enter in a crucial way; see [Kue91] for details on this important issue.

Chapter Nine

Extensions

9.1 OPERATOR SELFSIMILAR PROCESSES

The definition of selfsimilarity in $\mathbb{R}^d$ can be extended to allow for scaling by linear operators on $\mathbb{R}^d$ as follows [LahRoh82, HudMas82].

Definition 9.1.1 *An $\mathbb{R}^d$-valued stochastic process $\{X(t)\}$ is called "operator selfsimilar" if*
(a) it is stochastically continuous at each $t \geq 0$, and
(b) for every $a > 0$ there exists a linear operator $B(a)$ on $\mathbb{R}^d$ such that

$$\{X(at)\} \overset{d}{=} \{B(a)X(t)\}. \tag{9.1.1}$$

When D is a linear operator on $\mathbb{R}^d$, a^D will denote the linear operator

$$\exp\{(\log a)D\} = \sum_{j=0}^{\infty} \frac{1}{j!}(\log a)^j D^j$$

for $a > 0$. We say that an $\mathbb{R}^d$-valued stochastic process $\{X(t)\}$ is *proper* if for all $t > 0$, $\mathcal{L}(X(t))$ is full, namely it is not concentrated on any $d - 1$ dimensional hyperplane.

Theorem 9.1.1 [HudMas82] *Let $\{X(t), t \geq 0\}$ be a proper operator self-similar process. Then there exists an invertible linear operator D on $\mathbb{R}^d$ such that for each $a > 0$,*

$$\{X(at)\} \overset{d}{=} \{a^D X(t)\}. \tag{9.1.2}$$

The linear operator D in (9.1.2) will be referred to as an exponent for the process. The exponent D in Theorem 9.1.1 is not necessarily unique, see

[HudMas82]. Without the assumption of $\{X(t)\}$ being proper, the existence of a not necessarily invertible linear operator D is shown in Sato [Sat91].

Since the exponent is not necessarily unique in this case, we are interested in the set of all possible exponents D and uniqueness. Let $\text{End}(\mathbb{R}^d)$ be the set of all linear operators on $\mathbb{R}^d$ and $\text{Aut}(\mathbb{R}^d)$ the set of all invertible linear operators on $\mathbb{R}^d$. Define

$$\mathcal{E}(X(t)\}) = \left\{ D \in \text{End}(\mathbb{R}^d), \{X(at)\} \overset{d}{=} \{a^D X(t)\}, \forall a > 0 \right\}$$

and

$$S(\{X(t)\}) = \left\{ A \in \text{Aut}(\mathbb{R}^d), \{X(t)\} \overset{d}{=} \{AX(t)\} \right\}.$$

$S(\{X(t)\})$ is called the symmetry group of $\{X(t)\}$, and it is known to be a closed subgroup of $\text{Aut}(\mathbb{R}^d)$ (see [HudMas82]). For a closed subgroup H of $\text{Aut}(\mathbb{R}^d)$, the tangent space TH is defined by the set of all linear operators A on $\mathbb{R}^d$ such that $A = \lim_{n \to \infty} d_n^{-1}(D_n - I)$ for some sequence (D_n) in H and (d_n) in $(0, \infty)$, where $d_n \to 0$ and I is the identity operator on $\mathbb{R}^d$.

Theorem 9.1.2 [HudMas82] *Let $\{X(t)\}$ be proper and operator selfsimilar in $\mathbb{R}^d$.*

  (i) *For any $D_0 \in \mathcal{E}(\{X(t)\})$, $\mathcal{E}(\{X(t)\}) = D_0 + T(S(\{X(t)\}))$.*

  (ii) *$\{X(t)\}$ has exactly one exponent if and only if $S(\{X(t)\})$ is discrete.*

The next result links operator selfsimilar processes to the existence of a weak limit of scaled processes.

Theorem 9.1.3 [HudMas82] *Let $\{X(t), t \geq 0\}$ be a proper $\mathbb{R}^d$-valued stochastic process which is stochastically continuous. If there exist a stochastic process $\{Y(t), t \geq 0\}$ and $A(\lambda) \in \text{End}(\mathbb{R}^d)$ such that*

$$A(\lambda)Y(\lambda t) \overset{d}{\Rightarrow} X(t) \qquad \text{as } \lambda \to \infty, \tag{9.1.3}$$

then $\{X(t)\}$ is operator selfsimilar.

This corresponds to Theorem 2.1.1 (i). The following is an operator version of Theorem 2.1.1 (ii).

Theorem 9.1.4 [Mae98] *Let $\{X(t)\}$ be a proper operator selfsimilar process and let $D \in \mathcal{E}(\{X(t)\})$ be fixed. Then the norming operators $A(\lambda)$ in (9.1.3) can always be chosen so that, for each $s > 0$,*

$$A(\lambda s)A(\lambda)^{-1} \to s^{-D} \qquad \text{as } \lambda \to \infty. \tag{9.1.4}$$

Any other sequence of norming operators is of the form $G(\lambda)A(\lambda)$, where $G(\lambda) \in S(\{X(t)\})$.

Remark 9.1.1 (A historical remark on Theorem 9.1.4)   Laha and Rohatgi [LahRoh82] first extended the notion of *selfsimilarity* to *operator selfsimilarity* to allow for scaling by a class of linear operators on $\mathbb{R}^d$ and proved the results corresponding to Theorems 9.1.1 and 9.1.3. In their paper, they also proved that all norming operators satisfy (9.1.4). The difference between their results and the ones above is the following. As pointed out in [HudMas82], the scalings which are allowed in [LahRoh82] are only invertible positive-definite self-adjoint linear operators on $\mathbb{R}^d$. However, their family of processes is not closed under general affine transformations. Properties of positive-definite self-adjoint linear operators played an important role in their work. [HudMas82] claimed that $A(\lambda)$ in (9.1.3) varies regularly under some mild assumptions. (See [HudMas82], p. 283, lines 1–2 from the bottom].). However, a proof of this statement was not published until [Mae98].

9.2 SEMI-SELFSIMILAR PROCESSES

In this section, we introduce the notion of semi-selfsimilarity as an extension of selfsimilarity; see Definition 9.2.2. The latter definition will be motivated by examples, first in Property 9.2.1 on strictly semi-stable Lévy processes and second through the notion of diffusions on Sierpinski gaskets (Theorem 9.2.1).

Definition 9.2.1   *A probability measure μ on $\mathbb{R}^d$ is said to be strictly (α, a)-"semi-stable", if for some $a \in (0,1) \cup (1,\infty)$ and $\alpha \in (0,2)$, $\hat{\mu}(\theta)^a = \hat{\mu}(a^{1/\alpha}\theta)$, $\theta \in \mathbb{R}^d$.*

Property 9.2.1   *Let $\{Z_\alpha(t), t \geq 0\}$ be a Lévy process such that $\mathcal{L}(Z_\alpha(1))$ is strictly (α, a)-semi-stable. Then we have*

$$\{Z_\alpha(at)\} \overset{d}{=} \{a^{1/\alpha}Z_\alpha(t)\}. \tag{9.2.1}$$

Proof. Easy.   □

Once (9.2.1) is true for some $a \neq 1$, then it is also true for a^n, $n \in \mathbb{Z}$, replacing a. However, unless $\mathcal{L}(Z_\alpha(1))$ is stable, there exists $a > 0$ such that (9.2.1) does not hold, and thus it is not selfsimilar.

Kusuoka [Kus87], Goldstein [Gol87] and Barlow and Perkins [BarPer88] constructed diffusions on Sierpinski gaskets in the following way.

We consider $\mathbb{R}^2$. Let $a_0 = (0,0)$, $a_1 = (1,0)$ and $a_2 = (1/2, \sqrt{3}/2)$, and let

$F_0 = \{a_0, a_1, a_2\}$. Define inductively

$$F_{n+1} = F_n \cup \{2^n a_1 + F_n\} \cup \{2^n a_2 + F_n\}, \qquad n = 0, 1, 2, \ldots,$$

where $y + A = \{y + x, x \in A\}$. Let

$$G_0' = \bigcup_{n=0}^{\infty} F_n$$

and let G_0 be G_0' together with its reflection around the y-axis. Let

$$G_n = 2^{-n} G_0, \qquad n \in \mathbb{Z}, \qquad G_\infty = \bigcup_{n=0}^{\infty} G_n, \qquad G = \overline{G_\infty}.$$

The G is the Sierpinski gasket. Define a simple random walk on G_n as a G_n-valued Markov chain $\{Y_r, r = 0, 1, 2, \ldots\}$ with transition probabilities

$$P\{Y_{r+1} = y | Y_r = x\} = \begin{cases} \dfrac{1}{4}, & \text{if } y \in N_n(x), \\ 0, & \text{otherwise,} \end{cases}$$

where $N_n(x)$ are the four nearest points of G_n. Consider

$$X^{(n)}(t) = 2^{-n} Y_{[5^n t]}, \qquad t \geq 0, \qquad n = 0, 1, 2, \ldots.$$

Theorem 9.2.1 *The process $\{X^{(n)}(t)\}$ converges weakly to a process $\{X(t)\}$, and*

$$\{X(5^n t)\} \overset{d}{=} \{2^n X(t)\}, \qquad \forall n \in \mathbb{Z}. \tag{9.2.2}$$

Definition 9.2.2 *An $\mathbb{R}^d$-valued stochastic process $\{X(t), t \geq 0\}$ is said to be "semi-selfsimilar" if there exist $a \in (0, 1) \cup (1, \infty)$ and $b > 0$ such that*

$$\{X(at), t \geq 0\} \overset{d}{=} \{bX(t), t \geq 0\}. \tag{9.2.3}$$

The statements (ii) and (iii) in the following theorem correspond to Theorems 1.1.1 and 1.1.2, respectively.

Theorem 9.2.2 [MaeSat99] *Let $\{X(t), t \geq 0\}$ be an $\mathbb{R}^d$-valued, nontrivial, stochastically continuous, semi-selfsimilar process. Then the following statements are true:*

   (i) *Let Γ be the set of $a > 0$ such that there is $b > 0$ satisfying (9.2.3). Then $\Gamma \cap (1, \infty)$ is nonempty. Denote the infimum of $\Gamma \cap (1, \infty)$ by a_0.*

(a) *If $a_0 > 1$, then $\Gamma = \{a_0^n, n \in \mathbb{Z}\}$, and $\{X(t)\}$ is "not" selfsimilar.*

(b) *If $a_0 = 1$, then $\Gamma = (0, \infty)$, and $\{X(t)\}$ is selfsimilar.*

(ii) *There exists a unique $H \geq 0$ such that, if $a > 0$ and $b > 0$ satisfy (9.2.3), then $b = a^H$.*

(iii) *$H > 0$ if and only if $X(0) = 0$ almost surely. $H = 0$ if and only if $X(t) = X(0)$ almost surely.*

The real number H is called the *exponent* of the semi-selfsimilar process. We call $\{X(t)\}$ *H-semi-selfsimilar*.

Lemma 9.2.1. *Let $\{X(t), t \geq 0\}$ be an $\mathbb{R}^d$-valued process. If $a > 0$ satisfies (9.2.3) with some $b > 0$, then b is uniquely determined by a.*

Proof. Suppose that $\{X(at)\} \overset{d}{=} \{b_1 X(t)\} \overset{d}{=} \{b_2 X(t)\}$. Since $X(t)$ is nondegenerate, we have $b_1 = b_2$ by Lemma 1.1.1   □

Proof of Theorem 9.2.2. We first show the statement (i). By Lemma 9.2.1, b in (9.2.3) is uniquely determined by a. We thus denote $b = b(a)$. Let us examine the properties of the set Γ. By definition, Γ contains an element of $(0, 1) \cup (1, \infty)$. Obviously $1 \in \Gamma$ and $b(1) = 1$. If $a \in \Gamma$, then $a^{-1} \in \Gamma$ and $b(a^{-1}) = b(a)^{-1}$, because (9.2.3) is equivalent to

$$\left\{X(a^{-1}t), t \geq 0\right\} \overset{d}{=} \left\{b^{-1}X(t), t \geq 0\right\}.$$

Hence $\Gamma \cap (1, \infty)$ is nonempty. If a and a' are in Γ, then $aa' \in \Gamma$ and $b(aa') = b(a)b(a')$, because

$$\{X(aa't)\} \overset{d}{=} \{b(a)X(a't)\} \overset{d}{=} \{b(a)b(a')X(t)\}.$$

Suppose that $a_n \in \Gamma$, $n \geq 1$, and $a_n \to a$ with $0 < a < \infty$. Let us show that $a \in \Gamma$ and $b(a_n) \to b(a)$. Denote $b_n = b(a_n)$. Let b_∞ be a limit point of $\{b_n\}$ in $[0, \infty]$. For simplicity, a subsequence of $\{b_n\}$ approaching b_∞ is identified with $\{b_n\}$. Denote $\mu_t = \mathcal{L}(X(t))$. We have

$$\hat{\mu}_{a_n t}(\theta) = \hat{\mu}_t(b_n \theta), \qquad \forall \theta \in \mathbb{R}^d.$$

If $b_\infty = 0$, then, taking the limit, we get $\hat{\mu}_{at}(\theta) = \hat{\mu}_t(0) = 1$, which shows that $X(at)$ is degenerate for every t, contradicting the assumption of the nondegenerateness. Hence $b_\infty > 0$. It also follows that $b_\infty < \infty$. In fact, if $b_\infty = \infty$, then $b(a_n^{-1}) = b_n^{-1} \to 0$ with $a_n^{-1} \to a^{-1} > 0$, which contradicts the fact just shown. For each fixed t, there is an $\varepsilon > 0$ such that $\hat{\mu}_t(b_\infty \theta) \neq 0$ for $|\theta| \leq \varepsilon$. Thus we have $\{X(at)\} \overset{d}{=} \{b_\infty X(t)\}$. Therefore $a \in \Gamma$ and $b_\infty = b(a)$. This shows that the original sequence $\{b_n\}$ tends to $b(a)$. We denote the set of $\log a$ with $a \in \Gamma$ by $\log \Gamma$. Then, by the properties that we have proved, $\log \Gamma$

is a closed additive subgroup of $\mathbb{R}$ and $\log\Gamma \cap (0, \infty) \neq \emptyset$. Denote the infimum of $\log\Gamma \cap (0, \infty)$ by r_0. Then we have:

(1) If $r_0 > 0$, then $\log\Gamma = r_0\mathbb{Z} = \{r_0 n, n \in \mathbb{Z}\}$.

(2) If $r_0 = 0$, then $\log\Gamma = \mathbb{R}$.

To see (1), let $r_0 > 0$. Then, obviously, $r_0\mathbb{Z} \subset \log\Gamma$. If there is an $r \in \log\Gamma \setminus r_0\mathbb{Z}$, then $nr_0 < r < (n+1)r_0$ with some $n \in \mathbb{Z}$, and hence $r - nr_0 \in \log\Gamma$ and $0 < r - nr_0 < r_0$, which is a contradiction. To see (2), suppose that $r_0 = 0$ and that there is r in $\mathbb{R} \setminus \log\Gamma$. As $\log\Gamma$ is closed, we have that $(r - \varepsilon, r + \varepsilon) \subset \mathbb{R} \setminus \log\Gamma$ with some $\varepsilon > 0$. Choose $s \in \log\Gamma$ satisfying $0 < s < 2\varepsilon$. Then $r - \varepsilon < ns < r + \varepsilon$ for some $n \in \mathbb{Z}$, which is impossible. This shows (2). Letting $a_0 = e^{r_0}$, we see that the assertion (i) of the theorem is proved.

We claim the following.

(3) If $X(0) = 0$ almost surely, then $b(a) > 1$ for any $a \in \Gamma \cap (1, \infty)$.

(4) If $b(a) \neq 1$ for some $a \in \Gamma \cap (1, \infty)$, then $X(0) = 0$ almost surely.

(5) If $b(a) = 1$ for some $a \in \Gamma \cap (1, \infty)$, then $X(t) = X(0)$ almost surely.

To see (3), suppose that $b(a) \leq 1$ for some $a \in \Gamma \cap (1, \infty)$ and that $X(0) = 0$ almost surely. Fix t, then $\hat{\mu}_{a^n t}(\theta) = \hat{\mu}_t(b(a)^n \theta)$, and hence $\hat{\mu}_{a^n t}(b(a)^{-n}\theta) = \hat{\mu}_t(\theta)$ for every $n \in \mathbb{Z}$ and $\theta \in \mathbb{R}^d$. Since $X(0) = 0$ almost surely, we have $\hat{\mu}_{a^n t}(\theta) \to 1$ uniformly in θ on any compact set as $n \to -\infty$. Hence $\hat{\mu}_{a^n t}(b(a)^{-n}\theta) \to 1$ as $n \to -\infty$. It follows that $\hat{\mu}_t(\theta) = 1$, that is, $X(t)$ is degenerate. This contradicts the nondegenerateness and hence proves (3). To see (4), let $b(a) \neq 1$ for some $a \in \Gamma \cap (1, \infty)$ and note that $X(0) \overset{d}{=} b(a)^n X(0)$, this implies that $X(0) = 0$ almost surely To prove (5), note that, since $\{X(a^{-n}t)\} \overset{d}{=} \{X(t)\}$ by $b(a) = 1$, we have
$$P\{|X(t) - X(0)| > \varepsilon\} = P\{|X(a^{-n}t) - X(0)| > \varepsilon\} \to 0$$

as $n \to \infty$, and $X(t) = X(0)$ almost surely.

Now we prove the assertion (ii). It follows from (3) and (4) that $b(a) \geq 1$ for $a \in \Gamma \cap (1, \infty)$. Suppose $a_0 > 1$. Let $H = (\log b(a_0))/(\log a_0)$. Then $H \geq 0$. Any a in Γ is written as $a = a_0^n$ with $n \in \mathbb{Z}$. Hence $b(a) = b(a_0)^n$. It follows that $\log b(a) = n \log b(a_0) = nH \log a_0 = H \log a$, that is, $b(a) = a^H$. In case $a_0 = 1$, we have $\Gamma = (0, \infty)$ and there exists $H \geq 0$ satisfying $b(a) = a^H$, since $b(a)$ is continuous and satisfies $b(aa') = b(a)b(a')$.

The assertion (iii) is a consequence of (3), (4) and (5). This completes the proof of Theorem 9.2.2. □

An important application of Theorem 9.2.2 (i) is the following. Suppose one wants to check the selfsimilarity of a process. If one follows the definition of selfsimilarity, one has to check (1.1.1) for all $a > 0$. However, suppose one could show the relationship (1.1.1) only for, for instance, $a = 2$ and 3. Then by

Theorem 9.2.2 (i), the fact that 2, $3 \in \Gamma$ implies that $\Gamma = (0, \infty)$ since $\log 2 / \log 3$ is irrational, and thus one can conclude that $\{X(t)\}$ is selfsimilar. Therefore, we have the following.

Theorem 9.2.3 [MaeSatWat99] *Suppose that $\{X(t)\}$ is stochastically continuous. If $\{X(t)\}$ satisfies (1.1.1) for some a_1 and a_2 such that $\log a_1 / \log a_2$ is irrational, then $\{X(t)\}$ is selfsimilar.*

References

[AdlCamSam90] R.J. Adler, S. Cambanis and G. Samorodnitsky (1990): On stable Markov processes, *Stochast. Proc. Appl.* **34**, 1–17.

[Alb98] J.M.P. Albin (1998): On extremal theory for selfsimilar processes, *Ann. Prob.* **26**, 743–793.

[AloMazNua00] E. Alós, O. Mazet and D. Nualart (2000): Stochastic calculus with respect to fractional Brownian motion with Hurst parameter less than 1/2, *Stochast. Proc. Appl.* **86**, 121–139.

[Asm99] S. Asmussen (1999): *Stochastic Simulation with a View Towards Stochastic Processes*, book manuscript (for a preliminary version, see http://www.maphysto. dk/publications).

[AstLevTaq91] A. Astraukas, J.B. Levy and M.S. Taqqu (1991): The asymptotic dependence structure of the linear fractional Lévy motion, *Lithuanian Math. J.* **31** (1), 1–28.

[AvrTaq00] F. Avram and M.S. Taqqu (2000): Robustness of the *R/S* statistic for fractional stable noises, *Stat. Infer. Stochast. Proc.* **3**, 69–83.

[BadPol99] R. Badii and A. Politi (1999): *Complexity. Hierarchical Structures and Scaling in Physics*, Cambridge University Press, Cambridge.

[Bai96] R.T. Baillie (1996): Long memory processes and fractional integration in econometrics, *J. Econometrics* **73**, 5–59.

[BarPer88] M.T. Barlow and E.A. Perkins (1988): Brownian motion on the Sierpinski gasket, *Prob. Theory Relat. Fields* **79**, 543–623.

[Bar98] O.E. Barndorff-Nielsen (1998): Processes of normal inverse Gaussian type, *Finance Stochast.* **2**, 41–68.

[BarMikRes01] O.E. Barndorff-Nielsen, T. Mikosch and S.I. Resnick (Eds.) (2001): *Lévy Processes: Theory and Applications*, Birkhäuser, Basel.

[BarPer99] O.E. Barndorff-Nielsen and V. Pérez-Abreu (1999): Stationary and self-similar processes driven by Lévy processes, *Stochast. Proc. Appl.* **84**, 357–369.

[BarPra01] O.E. Barndorff-Nielsen and K. Prause (2001): Apparent scaling, *Finance Stochast.* **5**, 103–113.

[Ber94] J. Beran (1994): *Statistics for Long–Memory Processes*, Chapman and Hall, London.

[Ber69] S.M. Berman (1969): Harmonic analysis of local times and sample functions of Gaussian processes, *Trans. Am. Math. Soc.* **143**, 269–281.

[Ber96] J. Bertoin (1996): *Lévy Processes*, Cambridge University Press, Cambridge.

[BhaGupWay83] R.N. Bhattacharya, R.N. Gupta and E. Waymire (1983): The Hurst effect under trends, *J. Appl. Prob.* **20**, 649–662.

[Bic81] K. Bichteler (1981): Stochastic integration and L^2-theory of semimartingales, *Ann. Prob.* **9**, 49–89.

[BinGolTeu87] N.H. Bingham, C.H. Goldie and J.L. Teugels (1987): *Regular Variation*, Cambridge University Press, Cambridge.

[Bon82] L. Bondesson (1982): On simulation from infinitely divisible distributions, *Adv. Appl. Prob.* **14**, 355–369.

[BooSho78] S.A. Book and T.R. Shore (1978): On large intervals in the Csörgő-Révéz theorem on increments of a Wiener process, *Z. Wahrscheinlichkeitstheorie verw. Gebiete* **46**, 1–11.

[Boy64] E. Boyan (1964): Local times for a class of Markoff processes, *Illinois J. Math.* **8**, 19–39.

[BreMaj83] P. Breuer and P. Májòr (1983): Central limit theorem for non-linear functionals of Gaussian fields, *J. Multivar. Anal.* **13**, 425–441.

[BroDav91] P.J. Brockwell and R.A. Davies (1991): *Time Series. Theory and Methods*, Springer, Berlin.

[BurMAeWer95] K. Burnecki, M. Maejima and A. Weron (1995): The Lamperti transformation for selfsimilar processes, *Yokohama Math. J.* **44**, 25–42.

[CamMae89] S. Cambanis and M. Maejima (1989): Two classes of selfsimilar stable processes with stationary increments, *Stochast. Proc. Appl.* **32**, 305–329.

[CarCou00] P. Carmona and L. Coutin (2000): Intégrale stochastique pour le mouvement brownien fractionnaire, *C.R. Acad. Sci. Paris* **330**, 231–236.

[ChaMalStu76] J.M. Chambers, C.L. Mallows and B.W. Stuck (1976): A method for simulating stable random variables, *J. Am. Stat. Assoc.* **71**, 340–344.

[Che00a] P. Cheridito (2000): Arbitrage in fractional Brownian motion models, preprint, Department of Mathematics, ETH Zurich.

[Che00b] P. Cheridito (2000): Regularised fractional Brownian motion and option pricing, preprint, Department of Mathematics, ETH Zurich.

[Che01] P. Cheridito (2001): Mixed fractional Brownian motion, *Bernoulli* **7**, 913–934.

[Cox84] D.R. Cox (1984): Long-range dependence, A review. In: H.A. David and H.T. David (Eds.) *Statistics, An Appraisal*, Iowa State University Press, pp. 55–74.

[CsoRev79] M. Csörgő and P. Révész (1979): How big are the increments of a Wiener process? *Ann. Prob.* **7**, 731–737.

[CsoRev81] M. Csörgő and P. Révész (1981): *Strong Approximations in Probability and Statistics*, Academic Press, New York.

[Dacetal01] M.M. Dacorogna, R. Gençay, U.A. Müller, R.B. Olsen and O.V. Pictet (2001): *An Introduction to High Frequency Finance*, Academic Press, New York.

[Dah89] R. Dahlhaus (1989): Efficient parameter estimation for selfsimilar processes, *Ann. Stat.* **17**, 1749–1766.

[DaiHey96] W. Dai and C.C. Heyde (1996): Itô formula with respect to fractional Brownian motion and its application, *J. Appl. Math. Stochast. Anal.* **9**, 439–448.

[Dal99] R.C. Dalang (1999): Convergence of Markov chains to selfsimilar processes, preprint, Department of Mathematics, Ecole Polytechnique de Lausanne.

[DecUst99] L. Decreusefond and A.S. Üstünel (1999): Stochastic analysis of the fractional Brownian motion, *Potent. Anal.* **10**, 177–214.

[DelMey80] C. Dellacherie and P.A. Meyer (1980): *Probabilités et Potentiel*, Hermann, Paris.

[DieIno01] F.X. Diebold and A. Inoue (2001): Long memory and regime structuring, *J. Econometrics* **105**, 131–159.

[Dob79] R.L. Dobrushin (1979): Gaussian and their subordinated selfsimilar random generalized fields, *Ann. Prob.* **7**, 1–28.

[Dob80] R.L. Dobrushin (1980): Automodel generalized random fields and their renormalization group, in: R.L. Dobrushin and Ya.G. Sinai (Eds.) *Multicomponent Random Systems*, Marcel Dekker, New York, pp. 153–198.

[DobMaj79] R.L. Dobrushin and P. Májòr (1979): Non-central limit theorems for non-linear functionals of Gaussian fields, *Z. Wahrscheinlichkeitstheorie verw. Gebiete* **50**, 27–52.

[Doo42] J.L. Doob (1942): The Brownian movement and stochastic equations, *Ann. Math.* **43**, 351–369.

[DunHuPas00] T.E. Duncan, Y. Hu and B. Pasik-Duncan (2000): Stochastic calculus for fractional Brownian motion I. Theory, *SIAM J. Control Optim.* **38**, 582–612.

[ElN99] C. El-Nouty (1999): On the large increments of fractional Brownian motion, *Stat. Prob. Lett.* **41**, 167–178.

[EmbKluMik97] P. Embrechts, C. Klüppelberg and T. Mikosch (1997): *Modelling Extremal Events for Insurance and Finance*, Springer, Berlin.

[Get79] R.K. Getoor (1979): The Brownian escape process, *Ann. Prob.* **7**, 864–867.

[GirSur85] L. Giraitis and D. Surgailis (1985): CLT and other limit theorems for functionals of Gaussian processes, *Z. Wahrscheinlichkeitstheorie verw. Gebiete* **70**, 191–212.

[Gol87] S. Goldstein (1987): Random walks and diffusions on fractals, in: H. Kesten (Ed.) *Percolation Theory and Ergodic Theory of Infinite Particle Systems*, Springer, IMA *Vol. Math. Appl.* 8, pp. 121–128.

[Gra80] C.W. Granger (1980): Long memory relationships and the aggregation of dynamic models, *J. Econometrics* **14**, 227–238.

[GriNor96] G. Gripenberg and I. Norros (1996): On the prediction of fractional Brownian motion, *J. Appl. Prob.* **33**, 400–410.

[Guietal97] D.M. Guillaume, M.M. Dacorogna, R.R. Davé, U.A. Müller, R.B. Olsen, O.V. Pictet (1997): From the bird's eye to the microscope: a survey of new stylized facts of the intra-daily foreign exchange markets, *Finance Stochast.* **1**, 95–129.

[Ham87] F. Hampel (1987): Data analysis and selfsimilar processes, in: *Proc. 46th Session Int. Stat. Inst.*, Bulletin of the International Statistical Institute, Tokyo, Vol. 52, Book 4, pp. 235–254.

[Har82] C.D. Hardin Jr. (1982): On the spectral representation of symmetric stable processes, *J. Multivar. Anal.* **12**, 385–401.

[HeyYan97] C.C. Heyde and Y. Yang (1997): On defining long-range dependence, *J. Appl. Prob.* **34**, 939–944.

[HudMas82] W.N. Hudson and J.D. Mason (1982): Operator–selfsimilar processes in a finite-dimensional space, *Trans. Am. Math. Soc.* **273**, 281–297.

[Hur51] H.E. Hurst (1951): Long-term storage capacity of reservoirs, *Trans. Soc. Civil Eng.* **116**, 770–799.

[Ito51] K. Itô (1951): Multiple Wiener integral, *J. Math. Soc. Japan.* **3**, 157–164.

[JanWer94] A. Janicki and A. Weron (1994): *Simulation and Chaotic Behaviour of α-Stable Stochastic Processes*, Marcel Dekker, New York.

[Jen99] M.J. Jensen (1999): Using wavelets to obtain a consistent ordinary least squares estimator of the long-memory parameter, *J. Forecast.* **18**, 17–32.

[Jon75] G. Jona-Lasinio (1975): The renormalization group: a probabilistic view, *Nuovo Cimento* **26B**, 99–119.

[Jur97] Z. Jurek (1997): Selfdecomposability: an exception or a rule? *Ann. Univ. Mariae Curie–Sklodowska Sect. A* **51**, 93–107.

[KarShr91] I. Karatzas and S.E. Shreve (1991): *Brownian Motion and Stochastic Calculus*, 2nd ed., Springer, Berlin.

[KasKos97] Y. Kasahara and N. Kosugi (1997): A limit theorem for occupation times of fractional Brownian motion, *Stochast. Proc. Appl.* **67**, 161–175.

[KasMae88] Y. Kasahara and M. Maejima (1988): Weighted sums of i.i.d. random variables attracted to integrals of stable processes, *Prob. Theory Relat. Fields* **78**, 75–96.

[KasMaeVer88] Y. Kasahara, M. Maejima and W. Vervaat (1988): Log-fractional stable processes, *Stochast. Proc. Appl.* **30**, 329–339.

[KasOga99] Y. Kasahara and N. Ogawa (1999): A note on the local time of fractional Brownian motion, *J. Theoret. Prob.* **12**, 207–216.

[KawKon71] T. Kawada and N. Kôno (1971): A remark on nowhere differentiability of sample functions of Gaussian Processes, *Proc. Japan. Acad.* **47**, 932–934.

[KawKon73] T. Kawada and N. Kôno (1973): On the variation of Gaussian processes, in: *Proc. 2nd Japan–USSR Symp. Prob. Theory*, Springer, Lecture Notes Math. 330, pp. 176–192.

[KesSpi79] H. Kesten and F. Spitzer (1979): A limit theorem related to a new class of self similar processes, *Z. Wahrscheinlichkeitstheorie verw. Gebiete* **50**, 5–25.

[KhoShi00] D. Khosnevisan and Z. Shi (2000): Fast sets and points for fractional Brownian motion, in: J. Azéma, M. Émery, M. Ledoux and M. Yor (Eds.) *Séminaire de Probabilités XXXIV*, Springer Lecture Notes in Mathematics 1729, pp. 393–416.

[Knu84] D.E. Knuth (1984): An algorithm for Brownian zeroes, *Computing* **33**, 89–94.

[Kol40] A.N. Kolmogorov (1940): Wienersche Spiralen und einige andere interessante Kurven in Hilbertschen Raum, *C.R. (Doklady) Acad. Sci. USSR (NS)* **26**, 115–118.

[Kon84] N. Kôno (1984): Talk at a seminar on selfsimilar processes, Nagoya Institute of Technology, February, 1984.

[Kon96] N. Kôno (1996): Kallianpur–Robbins law for fractional Brownian motion, in: *Probability Theory and Mathematical Statistics. Proc. 7th Japan–Russia Symp. Prob. Math. Stat.* World Scientific, pp. 229–236.

[KonMae91a] N. Kôno and M. Maejima (1991): Selfsimilar stable processes with stationary increments, in: S. Cambanis, G. Samorodnitsky and M.S. Taqqu (Eds.) *Stable Processes and Related Topics*, Vol. 25 of *Progress in Probability*, Birkhäuser, Basel, pp. 275–295.

[KonMae91b] N. Kôno and M. Maejima (1991): Hölder continuity of sample paths of some selfsimilar stable processes, *Tokyo J. Math.* **14**, 93–100.

[Kue86] H. Kuensch (1986): Discrimination between monotonic trends and long-range dependence, *J. Appl. Prob.* **23**, 1025–1030.

[Kue91] H. Kuensch (1991): Dependence among observations: consequences and methods to deal with it, in: W. Stahel and S. Weisberg (Eds.) *Directions in Robust*

Statistics and Diagnostics, Part I, IMA Volumes in Mathematics and its Applications, Springer, Berlin, 131–140.

[Kus87] S. Kusuoka (1987): A diffusion process on a fractal, in: K. Itô and N. Ikeda (Eds.) *Probabilistic Methods in Mathematical Physics, Proc. Taniguchi Symp., Katata 1985*, Kinokuniya–North Holland, Amsterdam, pp. 251–274.

[LahRoh82] T.L. Laha and V.K. Rohatgi (1982): Operator selfsimilar stochastic processes in $\mathbb{R}_d$, *Stochast. Proc. Appl.* **12**, 73–84.

[Lam62] J.W. Lamperti (1962): Semi-stable processes, *Trans. Am. Math. Soc.* **104**, 62–78.

[LiMcL86] W.K. Li and A.I. McLeod (1986): Fractional time series modelling, *Biometrika* **73**, 217–221.

[Lin95] S.J. Lin (1995): Stochastic analysis of fractional Brownian motions, *Stochast. Stochast. Rep.* **55**, 121–140.

[LinLi97] S. Ling and W.K. Li (1997): On fractionally integrated autoregressive moving-average time series models with conditional heteroskedasticity, *J. Am. Stat. Assoc.* **92**, 1184–1194.

[LipShi89] R.Sh. Lipster and A.N. Shiryaev (1989): *Theory of Martingales*, Kluwer Academic Press, Dordrecht.

[Mae83] M. Maejima (1983): On a class of selfsimilar processes, *Z. Wahrscheinlichkeitstheorie verw. Gebiete* **62**, 235–245.

[Mae86] M. Maejima (1986): A remark on selfsimilar processes with stationary increments, *Can. J. Stat.* **14**, 81–82.

[Mae98] M. Maejima (1998): Norming operators for operator-selfsimilar processes, in: I. Karatzas, B.S. Rajput and M.S. Taqqu (Eds.) *Stochastic Processes and Related Topics, A Volume in Memory of Stamatis Cambanis, 1943–1995*, Birkhäuser, Basel, pp. 287–295.

[MaeMas94] M. Maejima and J.D. Mason (1994): Operator-selfsimilar stable processes, *Stochast. Proc. Appl.* **54**, 139–163.

[MaeSat99] M. Maejima and K. Sato (1999): Semi-selfsimilar processes, *J. Theoret. Prob.* **12**, 347–383.

[MaeSatWat99] M. Maejima, K. Sato and T. Watanabe (1999): Exponents of semi-selfsimilar processes, *Yokohama Math. J.* **47**, 93–102.

[MaeSatWat00] M. Maejima, K. Sato and T. Watanabe (2000): Distributions of self-similar and semi-selfsimilar processes with independent increments, *Stat. Prob. Lett.* **47**, 395–401.

[Maj81a] P. Májòr (1981): Limit theorems for non-linear functionals of Gaussian sequences, *Z. Wahrscheinlichkeitstheorie verw. Gebiete* **57**, 129–158.

[Maj81b] P. Májòr (1981): *Multiple Wiener–Itô Integrals*, Lecture Notes in Math. No. 849, Springer, Berlin.

[Man71] B.B. Mandelbrot (1971): A fast fractional Gaussian noise generator, *Water Resources Res.* **7**, 543–553.

[Man75] B.B. Mandelbrot (1975): Limit theorems on the self-normalized range for weakly and strongly dependent processes, *Z. Wahrscheinlichkeitstheorie verw. Gebiete* **31**, 271–285.

[Man97] B.B. Mandelbrot (1997): *Fractals and Scaling in Finance*, Springer, Berlin.

[Man99] B.B. Mandelbrot (1999): *Multifractals and 1/f Noise: Wild Self-Affinity in Physics (1963–1976)*, Springer, Berlin.

[Man01] B.B. Mandelbrot (2001): *Gaussian Self-Affinity and Fractals*, Springer, Berlin.

[ManVNe68] B.B. Mandelbrot and J.W. Van Ness (1968): Fractional Brownian motions, fractional noises and applications, *SIAM Rev.* **10**, 422–437.

[ManRacSam99] P. Mansfield, S.T. Rachev and G. Samorodnitsky (1999): Long strange segments of a stochastic process and long range dependence, preprint, available at www.orie.cornell.edu/~gennady/newtechreports.html.

[Mar70] G. Maruyama (1970): Infinitely divisible processes, *Theory Prob. Appl.* **15**, 1–22.

[Mar76] G. Maruyama (1976): Non-linear functionals of Gaussian stationary processes and their applications, Lecture Notes in Math., No. 550, pp. 375–378, Springer, Berlin.

[Mar80] G. Maruyama (1980): *Applications of Wiener Expansion to Limit Theorems*, Seminar on Probability, Vol. 49, Kakuritsuron Seminar (in Japanese).

[Mic98a] Z. Michna (1998): Ruin probabilities and first passage times for selfsimilar processes, PhD Thesis, Department of Mathematical Statistics, Lund University.

[Mic98b] Z. Michna (1998): Selfsimilar processes in collective risk theory, *J. Appl. Math. Stochast. Anal.* **11**, 429–448.

[Mic99] Z. Michna (1999): On tail probabilities and first passage times for fractional Brownian motion, *Math. Methods Oper. Res.* **49**, 335–354.

[MikNor00] T. Mikosch and R. Norvaiša (2000): Stochastic integral equations without probability, *Bernoulli* **6**, 401–434.

[MikSta00] T. Mikosch and C. Stărică (2000): Is it really long memory we see in financial returns? in: P. Embrechts (Ed.) *Extremes and Integrated Risk Management*, Risk Books, Risk Waters Group, pp. 149–168.

[MikSta01] T. Mikosch and C. Stărică (2001): Long range dependence effects and ARCH modelling, in: P. Doukhan, G. Oppenheim and M.S. Taqqu (Eds.) *Long Range Dependence*, Birkhäuser, Basel, in press.

[MikStr01] T. Mikosch and D. Straumann (2001): Whittle estimation in a heavy-tailed GARCH (1,1) model, preprint, available at www.math.ku.dk/~mikosch/preprint.html.

[Mor84] B.J.T. Morgan (1984): *Elements of Simulation*, Chapman and Hall, London.

[NorValVir99] I. Norros, E. Valkeila and J. Virtamo (1999): An elementary approach to a Girsanov formula and other analytical results on fractional Brownian motions, *Bernoulli* **5**, 571–587.

[NuzPoo00] C.J. Nuzman and H.V. Poor (2000): Linear estimation of selfsimilar processes via Lamperti's transformation, *J. Appl. Prob.* **37**, 429–452.

[OBrVer83] G.L. O'Brien and W. Vervaat (1983): Marginal distributions of selfsimilar processes with stationary increments, *Z. Wahrscheinlichkeitstheorie verw. Gebiete* **64**, 129–138.

[Ort89] J. Ortéga (1989): Upper classes for the increments of fractional Brownian motion, *Prob. Theory Relat. Fields* **80**, 365–379.

[PipTaq00] V. Pipiras and M.S. Taqqu (2000): Integration questions related to fractional Brownian motion, *Prob. Theory Relat. Fields* **118**, 251–291.

[PipTaq01] V. Pipiras and M.S. Taqqu (2001): Are classes of deterministic integrands for fractional Brownian motion on an internal complete? *Bernoulli* **7**, 873–897.

[Pri98] N. Privault (1998): Skorokhod stochastic integration with respect to non-adapted processes on Wiener space, *Stochast. Stochast. Rep.* **65**, 13–39.

[RacMit00] S.T. Rachev and S. Mittnik (2000): *Stable Paretian Models in Finance*, Wiley, New York.

[RacSam01] S.T. Rachev and G. Samorodnitsky (2001): Long strange segments in a long range dependent moving average, *Stochast. Proc. Appl.* **93**, 119–148.

[Rip87] B. Ripley (1987): *Stochastic Simulation*, Wiley, New York.

[Rob95] P.M. Robinson (1995): Log-periodogram regression for time series with long-range dependence, *Ann. Stat.* **23**, 443–473.

[Rog97] L.C.G. Rogers (1997): Arbitrage with fractional Brownian motion, *Math. Finance* **7**, 95–105.

[Ros56] M. Rosenblatt (1956): A central limit theorem and a strong mixing condition, *Proc. Natl. Acad. Sci. U.S.A.* **42**, 43–47.

[Ros61] M. Rosenblatt (1961): Independence and dependence, *Proc. 4th Berkeley Symp. Math. Stat. Prob.*, University of California Press, Berkeley, CA, pp. 411–443.

[Ros90] J. Rosiński (1990): On series representation of infinitely divisible random vectors, *Ann. Prob.* **18**, 405–430.

[Ros91] S.M. Ross (1991): *A Course in Simulation*, Macmillan, New York.

[RubMel98] R.Y. Rubinstein and B. Melamed (1998): *Classical and Modern Simulation*, Wiley, New York.

[SamTaq89] G. Samorodnitsky and M.S. Taqqu (1989): The various linear fractional Lévy motions, in: T.W. Anderson, K.B. Athreya and D.L. Iglehart (Eds.) *Probability, Statistics and Mathematics: Papers in Honor of Samuel Karlin*, Academic Press, New York, pp. 261–270.

[SamTaq94] G. Samorodnitsky and M.S. Taqqu (1994): *Stable Non-Gaussian Processes*, Chapman and Hall, London.

[Sat80] K. Sato (1980): Class L of multivariate distributions and its subclasses, *J. Multivar. Anal.* **10**, 207–232.

[Sat91] K. Sato (1991): Selfsimilar processes with independent increments, *Prob. Theory Relat. Fields* **89**, 285–300.

[Sat99] K. Sato (1999): *Lévy Processes and Infinitely Divisible Distributions*, Cambridge University Press, Cambridge.

[Sch70] M. Schider (1970): Some structure theorems for the symmetric stable laws, *Ann. Math. Stat.* **41**, 412–421.

[Shi98] A.N. Shiryaev (1998): On arbitrage and replication for fractal models, *Research Report No. 2, 1998*, MaPhySto, University of Aarhus.

[Sin76] Ya.G. Sinai (1976): Automodel probability distributions, *Theory Prob. Appl.* **21**, 63–80.

[Sin97] Ya.G. Sinai (1997): Distribution of the maximum of a fractional Brownian motion, *Russian Math. Surveys* **52**, 359–378.

[Tal95] M. Talagrand (1995): Hausdorff measure of trajectories of multiparameter fractional Brownian motion, *Ann. Prob.* **23**, 767–775.

[Tal98] M. Talagrand (1998): Multiple points of trajectories of multiparameter fractional Brownian motion, *Prob. Theory Relat. Fields* **112**, 545–563.

[Taq75] M.S. Taqqu (1975): Weak convergence to fractional Brownian motion and to the Rosenblatt process, *Z. Wahrscheinlichkeitstheorie verw. Gebiete* **31**, 287–302.

[Taq79] M.S. Taqqu (1979): Convergence of integrated processes of arbitrary Hermite rank, *Z. Wahrscheinlichkeitstheorie verw. Gebiete* **50**, 53–83.

[Taq81] M.S. Taqqu (1981): Selfsimilar processes and related ultraviolet and infrared catastrophes, in: *Random Fields: Rigorous Results in Statistical Mechanics and Quantum Field Theory*, Colloquia Mathematica Societatis Janos Bolya, Vol. 27, Book 2, pp. 1027–1096.

[Taq86] M.S. Taqqu (1986): A bibliographical guide to selfsimilar processes and long-range dependence, in: E. Eberlein and M.S. Taqqu (Eds.) *Dependence in Probability and Statistics*, Birkhäuser, Basel, pp. 137–162.

[TaqTev98] M.S. Taqqu and V. Teverovsky (1998): On estimating the intensity of long-range dependence in finite and infinite variance time series, in: R.J. Adler, R.E. Feldman and M.S. Taqqu (Eds.) *A Practical Guide to Heavy Tails. Statistical Techniques and Applications*, Birkhäuser, Basel, 177–217.

[TaqWol83] M.S. Taqqu and R. Wolpert (1983): Infinite variance selfsimilar processes subordinate to a Poisson measure, *Z. Wahrscheinlichkeitstheorie verw. Gebiete* **62**, 53–72.

[Ton90] H. Tong (1990): *Non-linear Time Series Models*, Oxford University Press, New York.

[Tsa97] C. Tsallis (1997): Lévy distributions, *Physics World*, July 1997, 42–45.

[Urb72] K. Urbanik (1972): Slowly varying sequences of random variables, *Bull. Acad. Polon. Sci. Sér. Sci. Math. Astron. Phys.* **20**, 679–682.

[Urb73] K. Urbanik (1973): Limit laws for sequences of normed sums satisfying some stability conditions, in: P.R. Krishnaiah (Ed.) *Multivariate Analysis – III*, Academic Press, New York, pp. 225–237.

[Ver85] W. Vervaat (1985): Sample path properties of selfsimilar processes with stationary increments, *Ann. Prob.* **13**, 1–27.

[Whi00] B. Whitcher (2000): Wavelet-based estimation for seasonal long-memory processes, preprint available at http://www.cgd.ucar.edu/~whitcher/papers/.

[Whi01] B. Whitcher (2001): Simulating Gaussian stationary processes with unbounded spectra, *J. Computat. Graph. Stat.* **10**, 112–134.

[Whi51] P. Whittle (1951): *Hypothesis Testing, in Time Series Analysis*, Almqvist och Wicksel.

[Whi53] P. Whittle (1953): Estimation and information in stationary time series, *Ark. Mat.* **2**, 423–434.

[WonLi00] C.S. Wong and W.K. Li (2000): On a mixture autoregressive model, *J. R. Stat. Soc. B* **62**, 95–115.

[Xia97] Y. Xiao (1997): Hausdorff measure of the graph of fractional Brownian motion, *Math. Proc. Cambridge Philos. Soc.* **122**, 565–576.

[Xia98] Y. Xiao (1998): Hausdorff-type measures of the sample paths of fractional Brownian motion, *Stochast. Proc. Appl.* **74**, 251–272.

Index

Adjusted range, 82

Black–Scholes SDE, 47
  fractional Black–Scholes, 47
  and regularization, 49, 50
Brownian motion, 4
  1/2-selfsimilar, 4
  fractional, 5
  simulation, 68

Codifference, 35
Correlogram, 85

Distribution
  α-stable, 9
  Cauchy, 9
  Gaussian, 9
  H-selfsimilar, 15
  Lévy, 9
  Lorentz, 9
  strictly stable, 9

Fractional Brownian motion, 5, 43
  and Itô formula, 48
  as Wiener integral, 6, 8, 50
  covariance, 5
  large increments, 54
  maximum, 51
  multiple points, 53
  non-semimartingale property, 45
  occupation time, 52
  sample path properties, 43, 63
    Hölder continuity, 43, 45
    nowhere bounded variation, 44
  simulation, 71
  stochastic integrals with respect to, 47
Fractional Gaussian noise, 16
  as fixed point of renormalization
    group, 16

Gaussian process, 27
Getoor's example, 60

H (exponent of selfsimilarity), 2
H-ss (H-selfsimilar)
  H-ss, ii (with independent
    increments), 57
  H-ss, si (with stationary increments),
    19
Harmonizable fractional stable motion
  as stochastic integral, 31
  selfsimilarity, 31
Hermite process, 23
Hölder continuity, 43

Kawazu's example, 61
Kesten–Spitzer process, 40
Kolmogorov's criterion, 43

Lamperti transformation, 11
Lamperti's theorem, 13
Lévy process, 9
  α-stable, 10
  sample path properties, 63
  simulation, 69
Limit theorems
  and selfdecomposability, 58
  and the Hermite process, 23, 25
  and the Rosenblatt process, 17
  central, 16
  Kesten–Spitzer, 41
  Lamperti, 13
  Májòr's, 27
  noncentral, 25
  Rosenblatt noncentral, 17
Linear fractional stable motion
  and long-range dependence, 36
  as stochastic integral, 30
  sample path properties, 63
  selfsimilarity, 30
  simulation, 79
Log-fractional stable motion, 34

Long-memory (see also long-range
  dependence), 84
Long-range dependence
  and codifference, 35
  and GARCH models, 90
  and non-linear time series models, 91
  and sample Allen variance, 37
  and scaling, 91
  and statistical estimation, 84
  meaning, 90
  of linear fractional stable motion, 36

Multifractality, xi

Operator selfsimilar process, 93
Ornstein–Uhlenbeck process
  stable stationary, 12
  stationary, 12

p-variation, 48

R/S-Statistic, 82
Random walk in random scenery, 40
Regular variation, 13
  and Lamperti's theorem, 13
  and slow variation, 13
  index of, 13
Regularization, 49, 50
Renormalization group, 15
  and fractional Gaussian noise, 16
Rosenblatt process, 17

$S\alpha S$ (symmetric α-stable, $0 < \alpha \le 2$),
  28
Sample Allen variance, 37
Sample path properties
  of fractional Brownian motion, 63
  of linear fractional stable motion, 63
  of log-fractional stable motion, 63
  of selfsimilar stable processes with
    stationary increments, 63
Self-affine process, xi
Selfdecomposability, 58
  and limit theorems, 58
  and selfsimilarity, 58

Selfsimilar process, 1
  α-stable Lévy, 10
  Brownian motion, 4
  exponent H, 2
  fractional Brownian motion, 5
  Kesten–Spitzer, 40, 41
    as limit process, 41
  operator, 93
  semi, 95
  simulation, 77
  strict stationarity, 11
  with independent increments, 57
    and selfdecomparability, 58, 59
    Gaussian and nonstationary
      increments, 62
    Getoor's example, 60
    Kawazu's example, 61
    Sato's theorem, 58
  with infinite variance, 29
  with stationary increments, 19
    and finite variances, 22
    long-range dependence, 21
    moment estimates, 19
    sample path properties in stable
      cases, 63
    symmetric stable, 28
Semi-selfsimilar process, 95
Simulation
  of α-stable processes, 70
  of fractional Brownian motion, 71
  of Lévy processes, 69
  of stochastic processes, 67
Slow variation, 13
  and regular variation, 13
Stability, 9
Stable measure, 9
Stable process, 27
  as stochastic integral, 29
  $S\alpha S$, 28
  simulation, 70
Statistical estimation, 81
  adjusted range, 82
  correlogram, 85
    least squares regression, 87
  maximum likelihood, 87
    Whittle estimator, 89
  spectral analysis, 87
  the R/S-statistic, 82

Stochastic integrals
  with respect to fractional Brownian
    motion, 47
Sub-Gaussian process, 33

Sub-stable process, 32

Whittle estimator, 89

www.ingramcontent.com/pod-product-compliance
Ingram Content Group UK Ltd.
Pitfield, Milton Keynes, MK11 3LW, UK
UKHW030039181224
452424UK00002BA/12